Coming Home
to
CHINA

Coming Home to CHINA

Yi-Fu Tuan

University of Minnesota Press

MINNEAPOLIS • LONDON

Frontispiece: Yi-Fu Tuan in front of his childhood home in Chongqing, China, 2005.

Published by the University of Minnesota Press
111 Third Avenue South, Suite 290
Minneapolis, MN 55401-2520
http://www.upress.umn.edu

Library of Congress Cataloging-in-Publication Data

Tuan, Yi-Fu, b.1930
  Coming home to China / Yi-Fu Tuan.
    p. cm.
  ISBN-13: 978-0-8166-4991-4 (hc : alk. paper)
  ISBN-13: 978-0-8166-4992-1 (pb : alk. paper)
  1. Tuan, Yi-Fu—Travel—China. 2. Geographers—United States—Biography.
3. China—Description and travel. I. Title.
  G69.T84A3 2007
  951.06092—dc22
  [B]   2006028172

Printed in the United States of America on acid-free paper

The University of Minnesota is an equal-opportunity educator and employer.

12 11 10 09 08 07   10 9 8 7 6 5 4 3 2 1

Contents

Preface

When are you going back to China for a visit?
I have been asked this question again and again,
especially during the past ten years. As time passes, an
increasing urgency enters the question. At my present
age of seventy-four years, it is already a chore to travel
long distances. Soon I will not have the stamina to visit
a country that, for all its economic boom, is still Third
World in some ways. So it is now—the year 2005—
or never. In 2004, I received an invitation from a
consortium of architects in Australia, China, and the
United States to attend and give a talk at an architectural
conference in Beijing in early June 2005. The conference
sponsors will pay my airfare and five days' accommodation
in a luxury hotel. That package in itself sounds inviting.
Making it not only inviting but plausible is that I may be
able to go there with my Chinese colleague and friend at
the University of Wisconsin–Madison, A-Xing Zhu. He
returns to China frequently for his research. If I go with
him, or with him and his family, I can leave many
tedious matters in his hands. Indeed, he repeatedly offers
to go with me, saying that would be no trouble, although
I fail to see how that can be true.

My two brothers, Tai-Fu and San-Fu, have both been
back to China several times. They seem to have enjoyed
the experience. So why my reluctance? One reason
is physical: I have always had a nervous stomach, and

travel invariably makes it worse. Nothing is more embarrassing and demoralizing than to have a stomach that compels one to dash to the toilet at an awkward time or makes one feel stuffed with filth that one is powerless to expel. Coughing, brought on by polluted air, can also be a serious problem for someone my age. And then there are the demands of coping with frantic traffic in the city and of the many steps, often slippery and without railings, in the country where tourist sites are located.

More important are the psychological reasons. I left China in 1941 at age ten and, except for a brief stopover in Shanghai on our way from the Philippines to England, I never returned to the land of my birth. Living, studying, and then teaching in Australia, England, Canada, and the United States meant that, over time, I steadily lost facility in my native language. My two brothers also left China at an early age, but, unlike me, they eventually married Chinese women and speak Chinese, or a mixture of Chinese and English, at home. I have remained single, and I have lived, over a period of thirty-five years, first in Minneapolis and then in Madison, Wisconsin, where the Chinese community was and is small. Another difference is that my two brothers are both physicists. In the sciences of physics and engineering, Chinese students and faculty make up a sizable minority. I, by contrast, am a humanist geographer, and few humanist scholars on American campuses are Asian. So I find myself speaking English all the time, even—out of laziness—when I am among Chinese colleagues and students.

I am struck by the irony that my two brothers, who as physicists needed only fluency in mathematics to rise to the top of their profession, are also competent in

Chinese. I, a humanist whose working tool *is* language, *natural* language, find myself hobbled by an increasing lack of facility in the one language that ought to matter more to me than any other. What will I feel—what will my audience in China feel—when I give my lectures in English?

I have doubts about my identity and where I truly belong. Others have doubts about my identity, too, for I am often asked where I consider my real home. "By and large, Earth," is my flippant answer. "But *where* on Earth?" This follow-up question assumes that a particular place must exist at which I am most comfortable and toward which I am able to form the deepest attachment. Certain landscapes do have a great appeal for me, one that goes beyond aesthetics—the desert, for example. I feel a psychological affinity with it and can see it as home. But if by home we mean a certain human warmth as well, then for me that is wherever I am totally at ease with the language. Perhaps more than most people, I need words and phrases powerful and subtle enough to bridge the chasm of isolated selves, thereby starting a true community and a true home. At a dinner party in China, I want to be able to say more than just "Pass the soy sauce" and "Yes, I have had a good trip."

Of course, I know that close-knit communities can come into being with just such commonplaces of exchange. One does not have to be original, or offer fresh takes on the world or on oneself, to establish intimate links with another. After all, the most intimate of all human exchanges—between a mother and her infant, and between a man and a woman in sexual congress—are conducted in "baby talk." Words are not necessary when the natural eloquence of body language takes charge. These special (indeed unique) eloquences of the body are not, however, available to me for reasons of physiology and temperament. Eloquences to which I do have access, at least potentially, are not gestures and stances but rather words and sentences. My anxiety mounts at the idea of going to a country—not just any country, but one in which I was born and learned the basic skills of being human—where I shall lack the verbal means to communicate with another person.

I moved from Minneapolis to Madison in 1983. I have now lived here for twenty-two years, by far the longest stretch I have been in any one place. I feel comfortable with my routines and have only one nagging worry, which is that changes in the future will all be unpleasant ones of old age: falling on ice and dislocating a spinal disk, losing eyesight and hearing, needing heart surgery or chemotherapy. Should I hang on to the present pleasantness for as long as I can, or risk doing something that could move physical disorders closer but that could also give me—if all goes well—one more noteworthy experience, another lease on life?

I opted for the possibility of another lease on life. In January 2005 I renewed my American passport, in March I applied for a visa, in April I purchased a plane ticket. I started the ball rolling. I was to leave Madison, with A-Xing, for Beijing on May 28 and return from Shanghai by myself on June 15. To show how anxious and neurotic I could be, a week or so before departure time I was sure I had a sore throat, which, from past experience, always prognosticated a bad cold that could make me helpless for the next three weeks. From one day to the next, I miserably checked the state of my throat rather than looked ahead to adventure in China.

The narrative of this trip, this journey into self and culture, has an unusual structure and style. First, it is not a diary, although it is based on notes I entered in a blank book two to three days after the event. This means that it is written mostly in the past tense; the tense changes to present to register that I am describing a state of mind that I not only held then but still hold. Second, a diary or journal is mostly a record of what one does, sees, and feels; it rarely records conversations and hardly ever elaborates ideas. I depart from this practice. The following journal takes me from one place to another, enlivened by brief descriptions, as would a travelogue, but unlike travelogues it includes the lectures I gave, as well as certain dinner conversations. I include them because they are occurrences on this trip. They happened! They show how I see the world and how I try to connect with people. They color my overall experience of China.

The Long Flight

7:15 a.m. Bob Sack came to pick me up. His offer to do so was comforting, for I never could trust the cab to arrive on time. The morning was clear and cool, which meant I didn't worry quite so much about bad weather holding up or—worse—canceling the flight. Bob, rather than dropping me off at the airport, came in to see me off. A-Xing was already there. Seeing him immediately lifted my spirit. He is a seasoned international traveler. He lives in Madison and teaches at the University of Wisconsin, but also has a research position at the Chinese Academy of Sciences in Beijing. I commute between home and office, a distance of half a mile. A-Xing commutes between Madison and Beijing, a distance of five thousand miles. Now, that's sophistication!

Age Difference

Our plane left the Dane County Airport on time. A-Xing and I were assigned separated seats. A man kindly offered to move so that we could be together. It's clear that this man and other people we met on our long journey, including flight attendants, saw us as related, probably father and son. Well, I have known A-Xing for eight years. I have got into the habit of seeing him as a colleague and friend, and as a bright star in the field of geographic information systems (GIS), but never as a "son." Nevertheless, on the eleven-hour flight from Minneapolis to Tokyo, A-Xing was filially solicitous. He taught me how to press the buttons to make the lounge chair curl into the right position, and, if I also wanted entertainment, how to raise the small screen attached to an armrest, call up a menu on the screen, and from it select the appropriate movie. Given my incompetence, the screen didn't seem so much to glow as to glower at me. Steeped thus in technology for the next few hours, when the steward announced dinner, I was almost expecting to be served a plate of assorted nutritional pills.

A Real Dinner

Meals in the business class, I was happy to find, still aspired to be an art form. I knew I was going to be fed well when the attendant offered me a hot towel, then put a white tablecloth on the small table that, with a bit of dexterous manipulation, emerged from the armrest. I had the choice of Western or Japanese cuisine. I chose Japanese. In appearance, the delicacies placed in front of me were Japanese art; in taste, they were the elixir of the gods, at least to a famished passenger used to the pretzels of economy flights.

Location Is Destiny

I am at an age when no place can seem friendlier than a clean, well-equipped, and, above all, unoccupied toilet. My pleasure was great when I found that my seat in the airplane was next to the toilets such that, without turning

my head, I could tell whether they were in use. I had
been forewarned that China, for all its new wealth and
sparkle, is still a backward country in hygiene: toilets,
even in high-class restaurants, are not the sort of place in
which one wants to read a magazine, and it is the better
part of valor to go there armed with a roll of toilet paper.

Beijing: First Impressions

Ejection from Childhood

We circled over Beijing for a half hour so that we could arrive on time at 9:25 p.m. As the plane hovered a few hundred feet above the runway, I looked out of the window and saw a dark night scene with scattered low-lying houses. That was the China I expected to find, for that was the China I had left behind. Still drowsy, I recalled the smell of manure and the noise of croaking frogs that dark night on the outskirts of Chongqing so long ago. The bump our airplane made as it touched ground woke me from my reverie. We passengers struggled up from our womblike seats, where we had slept, eaten, and burped like pampered babies, to stand on our own two feet. We even had to remove our suitcases from the overhead racks, which seemed unfair after all the mothering we had received. One of the shocks of long-distance air travel is this brutal ejection from childhood to adulthood. I speak for myself, of course, and for others like me. For A-Xing and people like him, there was no shock, since throughout the flight they had retained their adult standing by electronically communicating with the world thirty thousand feet below.

My first impression of the Beijing airport—a dark strip in the countryside—was dramatically overthrown when I entered the vaulting, bright space of the airport itself. A-Xing and I stepped onto an escalator. As soon as we got off at the first platform, which seemed to hover in midair, A-Xing took out his digital camera and snapped a picture of me, self-consciously reentering the country of my birth.

Speaking Chinese

The ultramodern architecture of the airport eased, at first, the feeling of strangeness. I could have been in Philadelphia or Paris. Then I became conscious of Chinese faces, announcements in Chinese over the intercom, and, more generally, the hum of Chinese speech. How bizarre to hear Peking dialect (Mandarin) spoken all around me rather than English, French, or, for that matter, Cantonese. Now, Cantonese wouldn't have seemed so strange, for it would be what I would hear if I had plunged into the busy streets of Chinatown in Chicago or New York.

But Mandarin! I realized there and then that I could not possibly make A-Xing and me the odd couple by speaking English, which was the language we used in Wisconsin. A-Xing must have been taken by surprise. It might not even have occurred to him that I could speak Chinese. Chatting with him in our native tongue as we walked down the hallways of the airport, I immediately discovered that we were using different forms of the pronoun "you." It was natural for me to use the familiar "you" (*ni*) with him, for, after all, we have known each other for many years. But he used the formal or polite *nin*. I was suddenly reminded of our yawning gap in age and also of the contrast in social practices: American laxness versus Chinese formality.

The Beijing Friendship Hotel

We were met at the terminal by Huang Juzheng, editor of the *Architect,* and his driver-assistant. (I was soon to realize that only expert drivers could survive Beijing's hectic traffic.) Huang studied architecture at Tung Chi University and earned his PhD in Japan. He was one of the organizers of the architectural conference that I had come to attend. A-Xing and I were led to a black sedan, an Audi. That took me by surprise. What did I expect? A less fancy and expensive car, but then I was used to the less well-heeled world of academics. I did expect the Beijing Friendship Hotel to be rather grand, and it was. From the outside it showcased traditional palace architecture, but inside it was all Hyatt, sporting marble-veneered floors and elevators that slid up and down in well-mannered hush. Differing from the Hyatts of the world, however, was its clear layout. One couldn't mistake front for back, left wing for right wing, at the Friendship Hotel. I appreciated this clarity after the long and disorienting flight.

A-Xing and Huang Juzheng accompanied me to my room, and that was a good thing, for without them I wouldn't have known how to turn on the lights. Like a simple American from the Corn Belt, I searched for switches, but there were none to be found. In advanced China, one turns on the lights by pressing the appropriate squares on a panel next to the bed.

A Walk in the Neighborhood and a Gustatory Shock

A-Xing and I thought we needed exercise. A morning walk around the hotel's neighborhood, we thought, would be good for us, even though the day, already turning warm and muggy, discouraged it. I was curious as to what I would see and how I would respond. I expected skyscrapers. Nevertheless I was jarred by their presence in a city that, fifty years ago, was a timeless cosmic diagram. A few of the buildings were ultramodern, by which I mean they aspired to lines and shapes that worked against intuition. For example, a building's flank might flare outward as it rose, making the entire structure look as though it might topple. Not suitable, from a psychological point of view, for a bank or financial center, I thought. Architects, thanks to the extraordinary power given them by engineering and materials science, could indulge their fantasies and produce sculptures to be seen and admired rather than places in which people could live or work in comfort and ease.

High-rises and an Urban Park

Most high-rises in central Beijing are not, of course, fanciful pieces of art, but are rather tacky tenement

9

houses for the lower middle and working class. They look like those residential towers that, in the United States, are judged failures. In Beijing, they are a social success. The government has built small landscaped enclosures for each cluster of tenements. In the United States, these could quickly become scenes of drug dealing and violent crime. But not in Beijing. A-Xing and I walked through several of these parks. They were heavily used. Men played cards, *wei chi* (the Japanese game Go), or mah-jongg, watched by other men. Children had their own playground with the usual array of brightly colored swings, slides, monkey bars, and merry-go-rounds absent painted horses or other refinements that might be damaged by weather and heedless young bodies. As I walked by, my ears were assaulted by a flood of sounds that in volume, pitch, and rhythm were indistinguishable from what I might hear at playgrounds anywhere in the world. The words were Chinese, but they might as well have been American English or Dutch. This made me think that if I ever needed to feel at home in a strange city, I could always go to a playground, shut my eyes, and sink into the roar of uninhibited young life.

The children's playground made me think of the universal. Next to it was a playground for oldsters that seemed to me unique to China. I could barely believe my eyes when I saw grandpas and grandmas twisting and turning on brightly colored poles that had handles and footrests, or bouncing happily on metal coils that were permanently fixed to the ground. Elderly Beijing residents did not have to confine themselves to their rooms. They could go outside, give their frail bodies a gentle workout on fixtures designed for them, and at the same time keep an eye on the children, not only for the children's sake but for their own pleasure.

"Silicon Street"

One of the places A-Xing and I visited on our walking tour was Beijing's "Silicon Street"—a long avenue packed with firms and shops that catered to the electronics industry and businesses. We entered one shop that, from the outside, looked like a supermarket or Wal-Mart, but inside was a maze of cubicles, each of which had a computer, a man or a woman, and display cases of what looked like cheap jewelry, but were in fact components of the computers that kept Beijing, China, and the world humming. How different the interior of this shop is from the interior of a body shop for motorcars, I thought. One would never mistake the body parts there for jewelry!

Cars link people by physically transporting them from one place to another. Computers allow people to stay put, while information moves around at lightning speed. On Beijing streets and intersections, traffic crawled—evidence that vehicular transportation and communication had advanced to a sclerotic stage. Fortunately, people in their stalled cars could use their cell phones to maintain contact. Cars and buses belonged to an older age. In a fanciful mood, I see them as elephants, jostling one another to reach the water hole. In striking contrast are the tiny cell phones and elegant laptops, their messages flowing unimpeded like birds through space.

"What's the Chinese word for 'computer'?" I asked A-Xing. *"Tien nao* [electric brain]," he replied. I had a

vision there and then of a future in which I could repair
or update my brain with jewel-like components
purchased at a store that would be far more advanced
than the one I visited in Silicon Street, a store that was
clean and slightly perfumed, like the cosmetics section of
a large department store. If a tired face could be repaired,
why not a tired brain?

Water-boiled Fish: A Delicacy

"For dinner," A-Xing said, "what do you say we try the
current food craze in Beijing, a dish called 'water-boiled
fish'?" I swallowed and agreed to try. (Actually, the fish
was boiled not in water but in an oily, heavily spiced
soup.) The restaurant had two floors, an upper one
decorated in the cool Western style, and a lower one
that was much larger, packed with tables and noisy
customers. We opted for the lower floor. I was
immediately struck by the youthfulness of the customers.
They couldn't have been older than thirty. So far as I
could see, I was the only bald head in a sea of fully
matted black hair. Moreover, females far outnumbered
males. A-Xing explained to me that the young women
were mostly college graduates and midlevel managers in
the city's booming businesses. They earned good salaries,
which they spent eating out with friends their age and
buying stylish clothes and cosmetics in competition with
one another, but also to attract men. Why save when the
men were expected to provide the dowry, and when,
after marriage, they would be another source of income
and security? A young woman could choose a special
dish in a row of restaurants, a designer handbag in a

parade of stores. Given the ratio of males to females
in China, she could also take her time to select a life
partner from a generous lineup of suitable men.

All the servers at the restaurant were young women,
dolled up and deferential, filling our teacups whenever
the level dropped a quarter inch below the rim. They were
trained by the management rather than in restaurant or
hotel schools. A-Xing and I witnessed one such training
session, a relentless military drill of perfect obedience
that took place on the sidewalk, outside the restaurant,
for all passersby to see.

Where were the male employees—the young men
and boys? They clustered around a water tank in the
backstage. We customers were encouraged to go there
and pick out a fish, see it scooped out of the water with
a net, and beaten to death with the whacks of an iron
bar. Young males were apparently well suited to this
grisly task. Unlike the prettily dressed and coiffed girls,
the boys wore long dark overalls splashed with water and
caked with dirt. They had spiked or unruly hair. They
chatted among themselves and seemed to enjoy what
they did. Yin and yang were reversed here: on one side
was the brightly lit "yang" world of pretty serving girls;
on the other, the dark, death-dealing "yin" world of men
and boys.

A-Xing and I went back to our table. We nibbled
cold delicacies and sipped chrysanthemum tea to pass
the time. Then a large brass bowl arrived. Floating on
top of the boiling soup were pieces of white flesh. The
transition from life to death was too sudden for me.
I could still see the fish twisting and turning violently
under the boy's muscular hand; I could still see it

thrashing its tail back and forth even as it was dealt
frenzied blows; and I could see it finally dead. But for it
to be not only dead but turned into chunks of meat
made me gag. I left the table to go to the washroom.
On my way there, my thoughts turned to whether the
washroom would stink of urine and feces, and whether
it was equipped with toilet paper.

Toilets and Civilization

Answer: no, it did not stink, and, yes, it did have toilet
paper. But the standard of cleanliness was far below that
of any halfway decent restaurant in North America. Why
is this so? I wondered. Why didn't Chinese civilization
make a greater effort to cosmeticize this least aesthetic
aspect of biological life? Europe did not do better until
the invention and wide use of the water closet and a
movement that linked health with hygiene in the
nineteenth century. And how was it that in the United
States, rather than in Europe, the washroom or bathroom
(note the euphemisms) became both a temple dedicated
to the belief that "cleanliness is next to godliness" and a
marbled chamber of self-indulgent luxury? Might the
Chinese tolerance for feces and stench be a consequence
of the use of manure on farmlands, producing a sour
fragrance that was pervasive, and the association of that
odor with fertility? Americans chose to repress all evidence
of decay and death. Chinese, by contrast, didn't feel called
to take that step because, to them, decay and its product
(manure) were a prelude to plant growth and life.

The Summer Palace

A university car came at 8:30 a.m. to take A-Xing and me to the Summer Palace. The Summer Palace dates back to the twelfth century and assumed roughly its present configuration in 1749. British and French troops destroyed it in 1860. In 1886, the Dowager Empress Cixi (1861–1908) rebuilt the garden and gave it its present name, Yeheyuan—Garden of the Preservation of Harmony. Foreign troops destroyed it one more time in 1900. Again, Cixi had it rebuilt, using money that was originally designated for the navy. Preservation of Harmony? There was precious little human harmony in the garden's modern existence.

A Solicitous Guide: Mrs. Kao

As soon as A-Xing and I entered the gate, we were accosted by unofficial and illegal guides. One middle-aged woman was most persistent. We accepted her service after reducing her asking price of thirty yuan to twenty yuan, and also because we thought she might be useful taking our pictures. Our guide (who went by the name of Mrs. Kao) addressed me as "aged sir" and told me that her father was about my age. As we strolled the grounds, Mrs. Kao became more and more friendly, even familiar. She told me again and again to sit up straight, to stand and walk without slouching, and that I must drink at least one glass of milk a day. In a confidential aside to A-Xing loud enough for me to overhear, she said that old people were like children and that one must keep an eye on them. Talk about cultural differences! I can't imagine an American guide exhibiting such personal solicitousness, or seeking my favor by telling me that I am well over the hill. How did I respond? After all, I have lived in the United States for more than fifty years and have become thoroughly acclimatized to its values. Well, there remains in me an unsuspected Chinese core, for I liked her familiarity—her constant reference to my age—and so, at the end of the tour, I gave her twenty-five yuan.

Overwhelmed by Symbols

We walked the Long Gallery (said to be the longest in the world), which framed the northern shore of Kunming Lake, to the Marble Boat, and then took a motorized craft, brightly colored and decorated to make it look "traditional," back to near the gate where we had entered. Throughout the tour, Mrs. Kao regaled us with her knowledge of the history of the garden and the rich symbolism of its plan, buildings, and sculptures. I listened to her while my eyes swept over the scene. The result was a baffling "cognitive dissonance" (as psychologists call it) between what I heard, what I knew of the garden's past, and what I saw. Mrs. Kao intended to entertain and astonish me with the splendors of the place, but she succeeded only in making me feel deeply saddened by the follies of the past: the use of money desperately needed to defend the nation to build a fantasy world for the amusement of essentially one individual—the Dowager Empress Cixi, her folly culminating in the construction of the Marble Boat, designed to look like a Mississippi paddle-steamer. On it she took tea.

As for the symbolisms of the site and the plan of the temples, pavilions, bridges, and sculptures, they all indicated, to my skeptical modern eyes, a fear of offending nature; and it would seem that the slightest offense, even when it was unintentional, such as a door opening at a wrong angle, a misplaced bronze lion, a color in an inappropriate shade, a bed that was insufficiently long (a long bed meant a long life), and so on, could bring misfortune and death to human occupants.

Appreciating with the Senses

But that was what I heard and remembered. What spread before my eyes was altogether sunnier—a typical landscaped garden (atypical only for its size) in which hordes of Chinese strolled, sat in pavilions, paddled on the lake, and, in one way or another, relaxed and enjoyed themselves. Nature seemed to me utterly benign that balmy Tuesday morning, and so did society, if only because I saw that society, like nature, could change for the better: under the last emperor, Pu Yi, people were allowed to enter the garden for a fee, and in 1924 it was opened to the public. Scenes of men, women, and children having a good time in the midst of beauty made me wonder whether the Chinese are better prepared to appreciate nature than Westerners because their sense of time, being less rigidly defined, allows them to be more relaxed.

To the European or American, a visit to the garden
or park is a one-time affair rather than an undertaking that
is broken into segments and spread over the seasons. To
the question, "Have you been to Yeheyuan?" the correct
Chinese answer is, "Yes. I have seen the cherry and plum
blossoms in late winter, wisteria and peonies in spring,
lotuses on the lake in August, and chrysanthemums
in autumn." In that sense, I have not really been to
Yeheyuan. Nor have I been, in another sense, captured
in a story told by the French writer, Simone de Beauvoir,
who spent an afternoon at the Summer Palace in 1955:
"In the middle of the lake I see a little boat: in it a young
woman is lying down peacefully asleep while two youngsters are frisking about and playing with the oars. Our
boatman cups his hands. 'Hey!' he calls. 'Look out for
those kids!' The woman rubs her eyes, she smiles, picks
up the oars, and shows the children how they work."

The First Thrill of Touch

A-Xing and I went back to the Friendship Hotel for
lunch. That afternoon, the Architectural Forum put up
an exhibition in a building across the street from the
hotel. Participants of the conference on architecture—
and I was one of them—were encouraged to attend if
only because the architectural firms, businesses, and
insurance companies provided us with financial support.
I walked around the exhibits and chatted with the
exhibitors, carrying more and more catalogs that were
kindly given to me, feeling more and more the pull of
gravity on flesh and bone, until one conference organizer
(a Mr. Shi), noticing my fatigue, asked a young fellow to

escort me back to the hotel. An escort? I didn't think
I needed one until he and I reached the curb. From there
I could see the hotel, but separating us from it was a
broad and unbroken stream of cars, buses, and bicycles,
none of which took the slightest notice of pedestrians
who wished to cross to the other side. The young man
took me by the arm and agilely threaded us through the
traffic. We reached the hotel's revolving door. I thought
he would leave me there, but no, he accompanied me
into the lobby, and then right to the elevator. Only
when the elevator arrived and its doors slid open did he
consider his duty completed.

A Speech to Architects:
A Tour de Force?

One reason I decided to go back to China was
an invitation to speak at an architectural
conference. Its theme was "Topophilia and Topophobia,"
a catchy title under which architects and planners could
explore the extraordinary transformation of cities such as
Beijing and Shanghai from traditional to modern and, in
the past ten years, to postmodern. I was invited, I believe,
because two of my early books, *Topophilia* (1974) and
Space and Place (1977), became fairly influential in the
training of planners, architects, and landscape architects.
I was even credited with the word "topophilia," an
honor I do not deserve, for it was used earlier by the
Anglo-American poet W. H. Auden and the French
essayist Gaston Bachelard. When I first heard about
the conference, I explained that I was too old for the
long flight. "Well," the answer came, "you will not be
the oldest at the conference." My resistance finally broke
down when I realized that if I did not go this time,
I would never see China again.

At the conference in Beijing, I gave my speech in
English. It was translated simultaneously into Chinese,
as were the speeches given by other foreigners. I,
however, wasn't a foreigner. Being introduced by a
Chinese professor of architecture (in Chinese) as a

scholar of human geography, and then standing up and confessing that my Chinese wasn't up to the challenge made me feel self-conscious. But this admission didn't seem to faze the audience, to whom I could seem a curiosity: a Chinese who was not quite Chinese, an establishment figure yet also a maverick.

What did I say? I wanted to relate architecture to basic human experiences of space and time, and to encourage my listeners to reflect on the enormous changes in China's built environment, which, I believed, were possible if I did not clutter my presentation with too many details and shades of meaning. On June 1, 2005, I said:

Space, Time, and Architecture

We live in place, move through space, and we are temporal beings, by which I mean not only that we live in time, as of course all living things do, but that we humans are naggingly conscious of time. What does this entail, experientially, in a built environment? How do the arrangements in a built environment, from those in an individual house to those in the city, reflect and enforce our time consciousness, whether this be the diurnal and seasonal cycles, the stages of a human life, or directional and progressive as in much of modern experience and thought?

Awareness that time ceaselessly propels us forward makes us yearn for stability—for time to stand still so that we can make sense of and savor what is around us. The spiritually inclined yearn for more—a sense of the eternal. Religion and its architectural expressions have historically catered to these yearnings. This is no longer quite the case. Religion as a calling that engages the best minds has retreated before modern secularism. As for architecture, in booming cities such as Beijing and Shanghai, buildings sprout up that, in their metallic brightness, seem designed to defy time and corruption. Yet they may turn out to be disconcertingly mortal, for, unlike great buildings of the past, they are highly sensitive to aesthetic fashion and to innovations in technology.

Dizzying Change

A major civilizational shift in the last hundred years is the disappearance of almost all agricultural activities from the city. Even in the nineteenth century, in rapidly industrializing England, no city had severed itself from rural ties—ties that were exhibited various ways, including extensive sheep and cattle folds, and farming in the backyard. As for China, I can draw on my own experience of approaching a city wall and entering a gate, expecting to find bustling urban life and encountering farms and villages instead. Walled cities in China, in my childhood, contained much agricultural land. This now makes sense to me, for the traditional Chinese city, being a cosmos, necessarily included not only built-up areas but also their ultimate means of support—farming.

Farming weds us to nature. Toward it we feel, however, a certain ambivalence. To take the positive attitude first, a recurrent question, even when we live in a consumer's paradise, is this: What better life is there, in a deep sense, than to live close to the earth with its intoxicating odors, sounds, and sights? What greater sensual satisfaction is there than cool water to quench thirst, wholesome food to relieve hunger, and a couch on which to slump in well-earned sleep? Both in the East and in the West, the elites of society have praised life on the farm, and they did so not only because it was the material basis of their wealth and power; for, even if gentlemen farmers didn't actually put their shoulders to the plow, they couldn't help knowing from day-to-day engagements with farm life that it offered the keenest sensory rewards.

Against this idyllic image is the dark side of backbreaking labor and uncertainty. Nature may be fertile, but the fertility has to be coaxed—it requires labor. And nature is Janus-faced in that at any time its smile can turn into frown, its serenity into rage.

For this reason, it is personalized, treated as a being whose unruly power is tempered by a certain responsiveness to human deference and self-abasement, expressed in prayers and offerings. If by topophilia we have these gestures of deference in mind, the conclusion is unavoidable that its psychological base is actually topophobia— the *dread* of nature. To the question, What is more deeply rooted in human experience? the answer might well be anxiety and fear rather than delight and appreciation.

The Cosmic City

A landscape of farms is an achievement, an affirmation of our need for order and predictability. But it is not yet an architectural achievement, for architecture has come to mean something more formal and ambitious; it proclaims, with a touch of hubris, that humans aren't totally earthbound. Inspiration for architecture is, historically, heaven and its stars—their grandeur and regularity. Anxiousness and uncertainty, which are an inescapable part of being human, can be assuaged by building a cosmic city, anchored on the North Star, with walls aligned to the cardinal points. Cities of this type were fairly common in various parts of the ancient world. In China, the earliest example dates back to the second millennium BC, and among the most recent is Beijing, a city that retained its cosmic template until the middle of the twentieth century.

Chinese ritual books prescribe that the city should be rectangular and properly oriented, that its wall be pierced by twelve gates to represent the twelve months, that it should have an inner enclosed square to contain the royal residences, from which is to emanate the axial, north-south avenue, with a royal ancestral temple on one side and an altar of the earth on the other. So conceived, the city is clearly not just a plan but also a timepiece that registers the daily and yearly courses of the sun.

The Diurnal Cycle

The sun's daily course dictates periods of wakefulness and sleep, activity and rest. Awake, we follow a countless number of circular paths, from tiny ones in the house and workplace, to larger ones from the house to the workplace, returning in each case, as the sun sets, to the point of departure that is called home. All these paths occur within the city. To follow a path beyond it for any distance would break the daily round that begins and ends in the home, and so offers a different kind of experience—the experience of the traveler.

Within the city, each pause in the circular path is a place, be it small, such as the bed we sleep in and the workbench we sit at, or larger, such as rooms built for various purposes, or larger still, such as a neighborhood, a marketplace, a street. To most of us, home is a special place, and it is so for a number of reasons, including the fact that it anchors the round-trip; also, home is where we spend the most time, experiencing it—unlike other places—during a good part of both day and night; and, importantly, home is where we rest to regain the energy on which all activities depend.

These small round-trips become routine and so barely register. Nor do we take particular notice of the places at which we pause during each trip; they too become routine and fade from memory. Much of life is reassuringly repetitious and unmarked. Even in the traditional walled city, the daily cycle and its routines are not marked by any special ceremony, a major exception being perhaps the opening and closing of the city gates, the one initiating social and economic life, the other reentry into the primordial world of darkness, sleep, and animal ease. The numerous practical tasks and routines of the day require, of course, an elaborate infrastructure of houses and streets, but these, however imposing in the aggregate, are seldom individually grand or distinctive. They may become so, however, when they are the setting for the larger seasonal and annual cycles, and when they commemorate special events.

The Seasonal Cycle

In an agricultural civilization, the seasonal cycle has a number of critical transitional moments that call for observation and ceremony, and these in turn may call for an appropriate stage. I have noted that the cosmic city is itself such a stage, its monumental walls so oriented that each side evokes a season: east is spring, south is summer, west is autumn, and north is winter. The cardinal points and the center are also in synesthetic and metaphoric correspondence to a host of other traits and qualities—in the Chinese case, the five elements, the five colors, the emblematic animals, human roles and functions. Together they make up a complex but orderly universe that is far different from the uncertainties people know on earth, whether they be caused by human passions and turmoil, or by the intemperances of nature.

Commerce and Merchants: The Marginalized

An orderly universe enjoys another advantage from the traditionalist point of view. It is well suited to a hierarchical social order of emperor, scholar-officials, farmers, artisans, and merchants. Farmers may be poor and occupy only a modest place in such an order, yet they command respect because, in a sense, the entire cosmos—the architecture and its rituals—is set up to ensure their way of life. Marginalized are people who do not own land and do not till the soil: they are the craftsmen and tinkers, and, above all, the traders and merchants, for they tend to be the most rootless, the most mobile. In the paradigmatic cosmic city, quarters are set aside for commerce behind the palace compound, on the north (the dark or yin) side of the world. Seldom, however, can this ritually appropriate location be made actual. Likewise, although officialdom makes every effort to regulate commerce and discourage it from spilling over to the residential compounds, this is almost never successful in a rapidly growing city.

As trade prospers, some merchants may become very wealthy, but their status remains modest. The only way for them to gain prestige is to acquire some of the trappings of officialdom, and these include Confucian manners and erudition, land ownership, and living in courtyard houses. The layout of these courtyard houses derives its symbolic import from an agricultural cosmology, and is therefore at odds with the merchant's own way of life. For all its large impact on the world, commerce fails to generate potent myths. Its engine and telos is money, and money is numbers, not the small ones of symbolic resonance but large ones—abstractions—that do not demand to be expressed in distinctive rituals and architecture.

Europe

The sketch I have given applies to China. Europe is different. Though it too is an agricultural civilization, its dependence on the cycles of nature and their symbols is complicated by a view of time that is historical and directional. I refer here to the Judeo-Christian view that time has a beginning, a middle, and an end, that the human story begins with creation and ends with consummation. In Europe, a conception of space anchored in symbolically potent cardinal points certainly existed: witness the post-Doric east-west orientation of Greek temples and the customary east-west alignment of Christian churches with the altar placed at the eastern end; witness also the rituals in the church calendar that were and are calibrated to accord with the passage of the seasons. But such manifestations of spatial cardinality in Europe are inconspicuous compared with Chinese space at all scales, from the walled city to sumptuous courtyard residences, to small farmhouses that open to the south. So, rather than pursue this line of thought further, I would like to turn to directional time—to time marked by a sense of past, present, and future.

Directional Time

An elemental sense of directional time is given by our body's asymmetry: the fact that it has a front and a back, and the fact that we have projects—that we live forward. What I see in front is the present and, possibly, the future. The back is the past, not visible to the eye and so, in this sense, it is also darkness. Such experience inclines us to assign certain social values to "front" and "back." It is, for example, rude to turn one's back on a person, since doing so makes him invisible and no longer of present concern. In most societies, a person's back and the space behind him are considered profane and dark, set aside to serve the body and other private needs. By contrast, his front and the space ahead are orderly and social, and may even be sacred. In China, at a macroscale, the emperor on his throne faces south, the noon sun, and the world of man. His back is turned toward darkness and a space prescribed for the transactions of commerce. At a micro-scale, the typical Chinese courtyard house has a well-defined front and back, each with its distinctive set of values. As for the Western world, the distinction between front and back—the one formal and presti-gious, the other informal and profane—is even sharper in the older public and private buildings. Guests enter through the front; servants and tradesmen through the back.

Directional time derived from the body is taken for granted, not something we dwell on or worry about. More intrusive on our consciousness and worrisome is direc-tional time that is given by our life's path from birth, through childhood and matu-rity, to death. This path and its stages are sometimes called a "cycle," but "cycle" is a misnomer, for old age, euphemistically labeled "second childhood," is not a return to childhood, but is rather the antecham-ber to death. Moreover, the end can come anytime, inflicted by hostile nature or human beings. Houses, for all their aes-thetic flourishes, are first and foremost shelters that protect us against these exter-nal threats. But our body is also nature, subject to pain, disease, and decay, and a source of anxiety. Against this threat, which attacks us from within, we respond with medicine, medical skills, and prayers; and we raise specialized buildings—hospices and hospitals, temples and churches—to accommodate them. All these measures are, however, temporary; they can postpone but not erase death.

Balms to Transiency

Transiency may be the human condition, but this doesn't mean that we are resigned to it. To the contrary, our nature impels us to seek consolations or balms. One balm is lineage: the idea that though an individual's life is short, lineage continues from remote ancestors to distant posterity. Another is to find permanence in such natural features as hills, valleys, forests, and the land itself. Their enduring presence—their restfulness—is a major reason for our deep appreciation of them. By comparison, a landscape of crops and houses does not last—or lasts only when it is laboriously maintained. If architecture has prestige, one reason lies in its air of permanence, projected by solid building materials and sheer size. If the edifice is also symmetrical, built in the form of a circle, square, or polygon, it is making a claim to being atemporal, eternal. Not having a front or back in itself suggests that the building is beyond the human need for orientation and movement, that it transcends the pettier divisions of time and their projects.

In an increasingly secular age, the well educated among us no longer find much consolation in religious doctrine and ritual, in sacred appurtenances and architecture, in lineage, or in a reputation so firm that it endures long after our demise. How, then, do we struggle against the fact that one day we die? We struggle with the help of technology, which enables us to prolong life, and also to make life so entertaining and fast-paced and so packed with bright objects, including architecture, that we can forget the inevitable end.

Technology: Extending Daylight and Height

One way to prolong life is to stay awake and engage with the world long after nightfall. But that is only practical on a broad scale with the invention of gaslight, and then electric light, at the end of the nineteenth century. I still find it surprising that, even in the 1850s, a great city's cultural and social activities must sharply diminish after dark, and that urbanites defer to nature's daily cycle almost as much as do country folk. Turning night into day is highly unnatural and may be counted among the most daring efforts to transcend our biological limitation. The success of the modern city in this regard is remarkable. After sunset, even the plainest streets and town squares turn, as though by magic, into rivers and islands of glittering light.

Distancing ourselves from nature may also be accomplished by elevation—by rising as far above the earth as is practical, and so deny our status as creatures that crawl at its surface, that eat and excrete, that depend on nature's abundance and our own sweaty labor. From the earliest times, height implied prestige. The taller the building, the more platforms it stood on, the more powerful was the message that man aspired to being godlike. There were exceptions, of course: courtyard houses in both Europe and China had greater prestige than high-rise tenement houses, the reason being that tall buildings were unsafe, liable to collapse, and their upper floors had no amenities. This remained true through much of the nineteenth century. In Paris, for example, the best apartments were on the first floor, immediately above the shops. There the bourgeois lived. At the next level, the rooms were smaller, the residents poorer, and so on until one reached the top floor, the attic, where indigent students and bohemians huddled in unheated and waterless rooms.

Within brief decades, extraordinary innovations in technology reversed the prestige of the floors. The higher one's living quarters are located, the greater is the prestige, until at the top of the building, where

the penthouse sits, we have—as the adver-
tising slogan goes—"the world at our feet."
Skyscrapers compete with one another in
height and elegance. They have sprung up
and continue to spring up at an ever more
frantic pace in periods of economic boom.
What do they mean? What do they say
about human aspiration? Or should I say,
human desperation? Or is it both?

Skyscrapers: What Do They Say?

Perhaps it has always been both. We human
beings want material sufficiency, if not
plenty. However, even material plenty is
seldom enough to silence the need in us to
escape decay, the ravage of time. In a small
but significant way, architecture answers
this need. The cosmic city is itself an outstanding example of a creation that transcends the routines of biological and social
life, that directs people's attention heaven-
ward. But both ritually and in architecture,
the city is still geared to the cycles of nature.
As for the Western world, until well into
the twentieth century, the silhouette of
a city is one of towering church spires.
Church spires point to the sky, reminders
that our destiny is not confined to this
world, that we follow a path—a directional
time—that begins with the earth but ends
in heaven. Inside the church, however,

services continue to cater to life's biological and social stages, and rituals are conducted to conform with nature's cycles.

In a modern city, glass-and-steel towers completely overshadow church spires. These new verticals barely recognize our standing as natural creatures. They deliberately cover up, as does the entire downtown area itself, evidences of downturn in the human cycle: injury and pain, corruption and death. Not only does a city dweller almost never see a human corpse, even dead squirrels and dead leaves are quickly removed from his eyes. In the hospital itself, shining surfaces of hygiene and efficiency hide the pathos of suffering and mortality. Few health and governmental buildings are, in fact, cloud-piercing skyscrapers. Work devoted to the body and to the running of government seems to require a feeling of stability and gravitas that is best projected by structures that hug the earth. Very tall buildings in today's skyline mostly cater to the needs of business and finance, which are directed to goals of abstract wealth that have no limit. And so an eye-catching architecture emerges in the landscape that, for the first time in the history of civilization, owes nothing to the template of a rotating cosmos.

Even people repelled by capitalism and great wealth must acknowledge that the tall edifices raised by famous architects in just the last decade are a feast for the eyes, a lift to the spirit. The rather rigid uprights and cylinders of the 1960s and 1970s, which can still evoke old-fashioned rectitude, may soon be outnumbered by skyscrapers that curve and gyrate in daring freedom. For the first time in history, architects feel free to disregard laypeople's intuitions of structural soundness and stability, or the need to adapt to either local topography or the stars in a readily discernible way. Some cities are beginning to look like gigantic sculptural gardens. The buildings in them are competing works of art, to be seen and admired, rather than places that mature into happy, eupeptic habitats. Do these daring buildings achieve greatness? Can there be greatness without a grounding in our biological nature, a clear reference to the cosmos, or a religious foundation?

Endurance: Buildings as Art

I have no answer except to say that one test for greatness in art is whether it endures. And here I have a troubling thought. In premodern times, buildings were not taken to be art. The more important ones— shrines, temples, churches, and the cosmic city itself—were religiously inspired. In the nineteenth century, commerce needed architectural showpieces to house its merchandise and, even more important, to advertise its newfound power and glory. World fairs and the buildings in them, such as the Crystal Palace and the Eiffel Tower, served that purpose. They were, however, vulnerable to criticism from an aesthetic point of view, as religious buildings of an earlier time were not. The Crystal Palace and the Eiffel Tower survived the criticism, and the Eiffel Tower stands today as a proud icon of French panache in design and a tourist attraction.

But that was the nineteenth century. In our time, criticism is more influential and potentially far more damaging. It is so in at least two ways: one, it has gained enormously in power by riding on the back of popular media, and, two, it is sensitive to fashion and technological innovation as never before. What if, after spending five hundred million dollars on a building, it is trounced as bland or kitschy by powerful critics, in the same way that a heavily financed film may be so trounced, ending in financial disaster? A half-empty movie palace eerily prognosticates the future of a skyscraper: new and only half-occupied, it may already show water stains and exude plugged-drain odor—signs of mortality beneath the bright sheaths of aluminum and glass.

Fashion is not solely a matter of aesthetic

taste; it is also a response to the pressure of advances in materials science and engineering. If, thanks to such advances, upright surfaces can bend, lean, or twist, so they will; moreover, tall buildings themselves may become dated as they are complemented, if not displaced, by spheres and Möbius strips. What this means is that monumental architecture no longer spells permanence—something that we humans, in our frailty and transiency, can latch on to and say, "Yes, but we have buildings of power and beauty that defy time."

A *Tour de Force?*

When my talk ended, I wondered whether it wasn't a little too abstract. The chair of our session, Professor Xing Ruan, probably had a moment of doubt, too. He thanked me and said, among other kind things, that my speech was a "tour de force." A tour de force is both an exceptional accomplishment and an accomplishment that is merely adroit or ingenious. That note of ambiguity is detectable in the evaluations that colleagues, both within and outside geography, have made in regard to my works. They seem entertained and even stimulated by my bold frames of reference—frames that they do not allow themselves for fear of committing naïvetés. I am granted, for reasons not clear to me, an unusual degree of freedom. In a long career, I have been asked to speak on topics of which I have no specialized knowledge: Chinese landscape painting to the Asia Society of New

York, aesthetics to the American Philosophical Society, folklore to the American Folklore Society, and, for that matter, architecture to an international gathering of architects. The indulgent treatment I receive from specialists who may be quite hard on one another makes me think that I serve as a sort of "wise fool" in the scholarly world. The effect on me is gratitude, but also the uncomfortable feeling that I am getting away with something and one day will be caught. But, then, that's how I feel about life itself, that I have lived this long by not being caught—yet—by a deadly virus or a drunken driver.

A Slightly Embarrassing Interview

I returned to my hotel to rest. At three o'clock, I went downstairs to meet with a young reporter from a magazine. She spoke very fast, using a rich vocabulary that made me feel I was being offered a literary treat. I struggled to understand and must have shown incomprehension on my face, for she tried again, speaking slower this time. Still I didn't understand enough to give her an appropriate reply. I sat in gloomy embarrassment. A painful and rather ridiculous image crossed my mind of an intelligent young woman trying to extract wisdom from an illiterate old man. She tried one more time. She asked me some simple factual questions concerning my early life. I was more successful answering her. Chinese words and phrases came out, at first hesitantly, then more copiously. I thought of myself as a ballpoint pen, too little used, that must be pressed hard against the paper for the ink to flow.

A Busy Day with
a Satisfactory Ending

Another Interview

The next morning, I met with another young female reporter in the hotel lobby. (I should slap my own face, Chinese style, for not remembering their names.) We chatted easily in Chinese. She told me that the average age of the magazine staff was only twenty-seven and that although the magazine was new, she and her fellow workers hoped it would soon be recognized as the Chinese *New Yorker*. Here is something I noted again and again on this trip—the energy and optimism of the young, their aspirations for themselves and for their country. What they aspired to was a universal standard, which to them was the Western standard (the best that the West had to offer in wealth, power, and knowledge) rather than anything specifically Chinese. So their magazine was to be a match for the *New Yorker*, their university a match for the best of the Ivy League, their national wealth a match for the wealth of the United States.

Forum on Urbanization

The forum that afternoon carried the rather grand title "Urbanization of China and the World." Most panel members were Chinese. Non-Chinese members came from the United States, Britain, and Australia. The discussion was lively and, at times, even slightly acrimonious. No one seemed shy. I certainly wasn't. In social gatherings, I tend to say little because I see no need to say more than just enough to launch another's pirouette into the limelight. Social friends don't listen, as we know from painful experience. Professional colleagues do, at least occasionally, for a major reason they attend meetings is to gather new information and knowledge.

Late in the afternoon, way beyond the scheduled closing time, the chair of the forum (Professor Yung-ho Chang of Harvard) asked the man to his right to make the concluding remarks. He did so because the man had been silent. The man said that he was a city planner, that he had wanted to break into the discussion several times but hadn't, not from shyness, but from compassion for his fellow panelists, who were all so eager to speak. What he wanted to say—and he welcomed the opportunity to say it now—was that he found much of the discussion too abstract to be useful. An anticlimax, I thought. I had the feeling that if the chair had offered the concluding remarks himself, it would have given everyone greater satisfaction. But then I recognized, once again, this weakness in me of always wanting an event, an undertaking, a day, a week, to come to a satisfactory close. Life just isn't like that, and maturity means accepting the loose ends, accepting the fact that conversations, more often than not, dribble to silence.

Dinner Treat by Young Architects

The conferees were going to a nearby restaurant for dinner and urged me to come along. I excused myself on the grounds that I was tired and that I needed to return to the hotel to rest. Now, that was something new. I hadn't always known that I could offer old age and its frailties as legitimate excuses. On my way back to the hotel, I heard footsteps behind me and a diffident voice calling my name. I turned and found four young men, who explained that they were architects in a large firm in the northeastern part of China, that they had come to Beijing for the conference, that they had attended my lecture and the forum discussion, and that they were interested in what I had to say. They asked whether they could join me for dinner. I couldn't bring myself to use the excuse of old-age fatigue one more time, so I said yes and suggested that we could all eat at the hotel.

The waitress handed a menu to each of us. My young guests consulted with me on my favorites and then selected one dish after another, all on the expensive side. I was somewhat surprised by their willingness to take charge. We chatted pleasantly on the events of the day, but soon the dreaded moment arrived when I realized that more was expected of me than comments on the weather. I struggled not to disappoint them. They listened politely. They too had to struggle to find the right words to loosen my tongue or, rather, to make it appear that I was not uttering inanities, that the things I said were worthy of their attention. An hour or so later, they observed that I must be weary after a long day and that they would leave so that I could rest. I was touched by their consideration and slightly taken aback by their style of speaking, which sounded to me cultivated, yet utterly sincere. Where, I wondered, did these young people pick up their manners? At the family dining table? In America, as everyone knows, the practice of the family eating together has gone out of fashion. It's good to know that in some things the Chinese still have a way to go before catching up.

When the check came, one of them grabbed it. "Wait a minute," I protested. "We are in China, and here custom dictates that I, being so much older than you, am to pay." They, for their part, insisted on doing the honors if only because I, a Chinese, had shown America what a Chinese intellectual could do. Even in the midst of my natural (though perhaps reprehensible) pleasure at their compliment, I couldn't help noticing that my young companions never doubted, despite my lack of facility in the language, that I was Chinese and that they themselves took great pride in being Chinese.

Another Hotel and a Campus Tour

"Sign on My Cap"

I packed in the morning in preparation for the move from the Friendship Hotel to the Jingshi Hotel at Beijing Normal University. The architectural conference was to end in the afternoon, and I had planned on listening to Xing Ruan's talk titled "Discerning the Good in China's Modern Architecture." The good. What the concept meant in life and, more narrowly, in architecture was a question of engrossing interest to me. Whereas I attended some presentations more from politeness than from curiosity, my desire to be at Ruan's was intellectually motivated. Unfortunately, the lecture was held up, and after forty-five minutes of waiting, I reluctantly left the room with A-Xing. In the hallway, students manned tables loaded with the usual assortment of information sheets and pamphlets. A young woman stopped me to ask for my autograph. A young man also wanted it, but he couldn't find paper for me to write on. After a minute or two of frantic searching, he took off his baseball cap and said, "Please, sign here!" I was handed a broad-nibbed pen, and I used it to write my name in bold Chinese characters. He put on the cap and grinned. I looked at his handsome face and yearned to possess him. After all, my name blazed on his forehead. But, sad to say, he was not for sale.

A-Xing's Alma Mater

When A-Xing told me that my next place of residence was to be at his alma mater, Beijing Normal University, I saw myself bedding down in a student dormitory and having to use the common bathroom at the other end of the hallway. So imagine my surprise when we drove into the resplendent courtyard of a luxury hotel—the Jingshi!

The hotel was built on university property and paid rent for the privilege, but in all other respects it was a business venture, and its success depended on attracting customers. Given its location, I imagined that a fair number of guests were visiting academics like myself, but most, I had to assume, were well-to-do businesspeople. The bellboy who opened the door, the maître d' who showed me to my table, the maid who cleaned my room, all addressed me as *lao shi* (teacher). I soon discovered that that was how they addressed every guest as a mark of respect. "Teacher" indicated status? Being *lao shi* commanded respect? That was a surprise. I knew, of course, that China traditionally revered the teacher, but I also thought that first communism, then capitalism, had done away with that frame of mind. When A-Xing wanted someone in the hotel or in a restaurant to show me special consideration, he might say, "Hey, this old gentleman is a scholar," and it worked. It worked the same way that it would have worked if, in the United States, I was introduced as the CEO of Kentucky Fried Chicken.

To pass the time before dinner, A-Xing showed me the campus. We strolled around the playing fields, and that somehow led me to memories of childhood in

Chongqing, when I used to walk around the playing fields of Nankai School before returning home for dinner. I hero-worshipped the athletes then. To a frail and sickly child, the older boys and girls who sprinted along the tracks and sailed over the hurdles were embodiments of physical perfection and prowess. Strange that I should feel the same awe, the same admiration, now that I have reached the other end of life.

But what happened in between, especially in the half century that I had spent in the United States? I had lost interest in sports totally, even as a spectator. The reason must be that on American campuses, athletics has become a special department, and sports (in particular, football and basketball) a quasi business activity that is conducted quite separately from day-to-day student life. On an American campus, I do not have the pleasure of walking by playing fields where I can see students taking pride in their physical fitness and ability, running, jumping, doing somersaults, or dribbling a ball. They do all these things in America, of course, but either indoors or in a space separate from the pathways of students and faculty going to and from classrooms and laboratories. Young Americans are turning into professionals, losing their standing as *student* athletes, losing the classical ideal of a perfect marriage between body and intellect, and losing thereby my interest.

Challenges of Translation

After the stroll around campus, A-Xing said we still had time before dinner and that we might spend it checking the Chinese translation of the lecture that I was to give the next day. We didn't find egregious errors, but we did find many infelicities. A-Xing called three students and asked whether they could come over to the hotel and help. These students were well known for their command of Chinese and also for their fair knowledge of English. They plunged into the high art of finding the right word and the right expression, and even the right rhythm, while I stood by feeling useless. We went out for dinner and then returned to the hotel for more work on the text. With characteristic thoughtfulness, they offered to continue with the translation in the hotel lobby so that I could go up to my room and rest.

Showing Off in Chinese and English

At Beijing Normal University

Xie Yun, dean of the College of Earth Science, and Zuo Yi-Ou, one of the student-translators from the evening before, escorted A-Xing and me to the auditorium, which I was told, sat around four hundred. We walked into a sea of faces—students and faculty from Beijing Normal, Bei Da, and Tsinghua—and, to my surprise, applause. In other parts of the world, only heads of state and rock stars could expect this kind of welcome. In China, throbbing with the dynamism of modernity and yet in some ways still traditional, it would appear that mere age wrapped in bloated reputation could still do the trick.

President Shih Pei Jun started the ceremony by bestowing on me the status of "adjunct professor" at his university. I had to think of a reply quickly and came up with something like this: "President Shih, I am not only honored but reassured by your bestowal of this status on me. Let me explain. Seven years ago, I retired from teaching. Ever since, I felt at a loss as to who I was and where I stood in society. When people asked, I didn't quite know what to say. As of this morning, I can say I am an adjunct professor of geography at your university."

Humanistic Geography

At the architectural conference, I had addressed a mixed audience of Chinese and Westerners, and so it seemed not altogether inappropriate to use English. But here I was speaking to an audience made up entirely of Chinese, so what was my excuse? I felt I had to explain. After the usual expression of thanks to my hosts, I said: "As you notice, I am speaking in English. I owe you an apology and a brief explanation. I left China with my family sixty-four years ago to live among English-speaking peoples—Australians, Brits, and, since 1951, Americans. Sad to say, in that time my Chinese has grown rusty. Though I can still carry on a social conversation in my native tongue, I am not able to do so on topics of intellectual interest. And this is doubly regrettable because my area of competence is humanistic geography. As I hope to show in the following lecture, if humanistic geography is distinguished in one characteristic, it is in its extraction of meaning from the resources of language. A humanist geographer who is not skilled in and sensitive to the subtleties of language is therefore a living contradiction."

The Felt Quality of Environment

I broached three themes in my lecture: the felt quality of environment, the psychology of power, and the relationship between material setting and the good life. By "felt quality," I refer to the obvious fact that places are not just what we see but also what we hear, smell, and touch, even taste. This felt quality is rich beyond description by virtue of the power of our five senses working alone, doubly, or, more often than not, all together. A more mysterious aspect of our senses is a phenomenon called synesthesia, a blending of the senses such that, for example, when one hears a sound one also sees a color. Generally, low-pitched sounds such as deep voices, drums, and thunder produce dark and round images, whereas high-pitched sounds such as soprano voices, violins, and squeaks produce bright and sharp images. Language points to its synesthetic grounding when we say in English, "What a *loud* tie you have" or "It's *bitterly* cold." What would be an example in Chinese? Could the expression *jin sang zi* (golden throat) be one?

Thanks to synesthesia, objects acquire a vividness and resonance they would not otherwise have. It is an advantage to young children because it helps them locate and fixate on the world's objects. When strongly developed, however, it promotes hallucination. As children grow older and acquire a certain fluency in language, synesthesia weakens, its function to enrich the world

being taken over by the metaphorical powers of language.

What is a metaphor? If synesthesia is the
blending of senses, metaphor is the blending of image-ideas or concepts. Metaphor
enables us to make concrete what is diffuse,
familiar what is unfamiliar. Nature, for
example, can seem diffuse, complex, and
threatening. It becomes less so when we
predicate it on parts of our body that we
know intimately. So we refer to headlands,
foothills, the mouth of a river, the spine of
a ridge, an arm of the sea, and so on. Even
the objects we have made ourselves can
seem coolly indifferent. To overcome that
detachment, we bind objects to our anat-
omy: the eye of a needle, the spine of a book,
the hands of a clock, the legs of a table, and
the back of a chair.

These are, of course, English idioms,
and I don't know that they all have Chinese equivalents. Some do exist, and some
are not only equivalent but identical. And
so we say in Chinese *ho kou* (river mouth)
and *shan chio* (mountain foot). A worthwhile project in humanistic geography is to
see how languages differ in the ways they
use metaphors to make unfamiliar objects
more familiar.

Not just metaphors but the full resources
of language are available to us as poets—
and we are all poets to some degree—to

firm up the emotional bonds between our-selves and the world. The world is made up of tangible objects, but also of more abstract entities such as space and spaciousness. How does language cope with spaciousness, making it more real and vivid to us? One way is to use the specialized vocabulary of numbers. For example, a popular medieval work (*South English Legendary*) conveys the vastness of space by saying, "If a man could travel upwards at the rate of more than 40 miles a day, he still would not have reached highest heaven in 8,000 years." But more common is to use a geographical vocabulary that can stimulate our geographical imagi-nation. I am struck by the similarity between two poems, one composed by an anony-mous Chinese poet in the Han dynasty and the other by the English poet Wordsworth in the nineteenth century. The Chinese poem, rendered into English by Robert Payne, has these lines: "Who knows when we shall meet again? The Hu horse leans into the north wind; the Yueh bird nests in southern branches." In Wordsworth's poem "The Solitary Reaper," just how solitary is the Reaper? How vast is the space that envelops her? For answer, Wordsworth, like the Chinese poet, calls up two con-trasting images: to one side are the "weary bands of travellers in some shady haunt, among Arabian sands," and, to the other, "the Cuckoo-bird, breaking the silence of the seas among the farthest Hebrides."

I have noted similarities and differences in the ways English and Chinese evoke the felt quality of environment. It may be that true synesthetic expressions ("it's bitterly cold") are more common in English than in Chinese, suggesting that the Chinese people have moved closer to depending on the purely linguistic devices of metaphor and simile. On the whole, educated Chinese are more apt to use poetic words and phrases in their speech and writing than are their European counterparts, one reason being that the Europeans, though not the Chinese, have at one stage of their development—the seventeenth century—denounced the use of adjectives and fancy metaphors. Members of the newly founded Royal Society did so in a fit of scientific passion. "Just the facts" turned into "just the numbers." Science, they believed, must eschew verbal embellishment. Within a society of any size, however, differences in attitude can be expected among individuals. Some take pride in unvarnished speech, others are less astringent, and still others may glory in the evocative powers of a rich vocabulary.

The Psychology of Power

My second theme is the psychology of power. Geographers are much concerned with the human transformation of the earth. Repeatedly, they seek to understand how forest and scrubland, steppe and swamp, have been turned into arable fields, towns, and cities. Neglected is the exercising of power for pleasure—the pleasure that is to be had in making gardens and pets. Geographers, like most people, tend to see gardens and pets as belonging to an area of innocence, in sharp contrast to large works of engineering and economic development, which are tarnished by suspicions of greed and pride. Nevertheless, from a psychological viewpoint, playing with nature, restrained only by the limit of one's fantasy, may manifest an even greater urge to power.

Consider water. Water becomes a pet when we make it dance for us. But we can only make it dance through the exercising of irresistible power—the power of hydraulic engineering and of large labor teams organized along military lines. The fountains that are the pride of the great gardens of Europe were mostly built by autocratic rulers in the seventeenth and eighteenth centuries. Many construction workers died in the effort, their bodies carried out in the middle of the night so as not to demoralize workers. Tourists who flock to these showpieces today, their senses drugged by beauty and charm, forget the raw power that lies behind them.

Water is alive only in a figurative sense. So let's move to things that are truly alive: plants, animals, and human beings. An egregious example of abusing nature for pleasure is the miniature garden—the *pen ching,* or, to use the more common Japanese word, bonsai. It is considered a fine art. But what kind of fine art is it that regularly uses instruments of torture—knives and scalpels, wires and wire cutters, trowels and tweezers, jacks and weights—to distort the plants and prevent their natural growth?

Animals are domesticated for economic use, but also made obedient through training or docile through breeding so they can be playthings and pets. Training can turn a huge and powerful animal, such as the elephant, into a docile beast of burden. Training can also turn it into a plaything—an object of ridicule—as when an elephant is made to wear a petticoat and stand on its hind legs. An even more radical way of altering nature is through selective breeding. Applied over successive generations, it can transform an animal into something dysfunctional and grotesque, and yet appeal to the taste of jaundiced connoisseurs. Think of the goldfish and the miniature dog—pets that have found favor in China. A certain breed of goldfish is created to have large eyes shaped like fish bowls that impede movement and are easily damaged.

As for the miniature dog, the Pekinese, weighing less than five pounds, is just a bundle of hair that can be used to warm the owner's lap.

From a psychological viewpoint, power reaches a peak, a peak charged with sadistic-erotic pleasure, when human beings themselves are turned into playthings. In Europe, Renaissance princes kept dwarfs, whom they dressed up, slobbered over, passed around at the dinner table, or presented as gifts to influential friends. Household slaves and servants, if they were comely, enjoyed the status of pets in slave-owning and other strongly hierarchical societies. In England, black boys were put in fancy uniforms so that they, along with purebred dogs, could sit for portraits with their masters and mistresses. Toward the end of the eighteenth century, so many duchesses and countesses kept black boys as pets that the fashion turned to Chinese or Indian boys. They were harder to come by and so commanded greater prestige. And then there are the women. In despotic Eastern societies, they were the decorative objects and sexual toys of powerful men, small, pretty, and helpless—a helplessness made evident in China when women submitted to having their feet bound and deformed. Even in relatively enlightened Western societies, women were legally children—child-wives in doll houses, as Ibsen put it—until a century or so ago.

Do I speak only of the past? Have times changed? The answer is yes, but the desire to dominate or patronize is too deeply embedded in human psychology to disappear altogether. Today, this desire is directed mostly at racial minorities in our own country, whether this be the United States or China, and at "our little brown brothers" in the rest of the world. More generally, it is directed at all subordinates. Dog owners like to order their dog to "fetch" and see the animal trotting off in obedience. But it is a pleasure available to all who have human subordinates at their disposal. The boss says "fetch"—though, of course, he uses a more polite language—and his subordinate goes to get coffee or a multimillion-dollar contract. Geographers have focused too exclusively on economic exploitation. As humanists, we should also attend to the ways we *toy* with nature and weaker people for no other purpose than to indulge in our dark fantasies of total power and control.

Environment and Quality of Life

My third theme is the relationship between the quality of environment and the quality of life. As swamps are drained and malaria is conquered, the quality of human life undoubtedly improves. Likewise in a built environment, as peeling walls are repainted, drains unclogged, and rooms and household amenities added. But at what point does adding more rooms and amenities cease to improve, and might indeed detract from, the quality of life—a life that is not only materially but intellectually and spiritually rewarding? China faces this question as its economy booms and standard of living rises so that people can move from shoddy dwellings to well-built ones, and (for some) even from well-built houses to luxurious mansions.

Material goods can enslave rather than liberate. This is well understood. But what about works of art? Don't they enrich? What about philosophy and religion? Don't they add to the quality of life? Take, first, the enlarging and enriching power of art; more specifically, of architecture. Consider an elemental aesthetic experience known to all human beings, that of interior space. The quality of that experience—of what it means to be inside and enclosed—varies enormously, depending on people's access to great works of architecture. Ancient Egyptians knew the sublimity of exterior space (think of the pyramids under moonlight), but interior space for them was darkness and clutter. Ancient Greeks had the Parthenon on the Acropolis to lift their spirits, but its interior was hardly more spacious than the interior of an Egyptian mortuary temple. Europeans had to wait for the construction of Hadrian's Pantheon in Rome (AD 118–28) to acquire, for the first time, the sense of an interior space that was formally elegant yet sublime, a vast hemisphere illuminated by the rotating sun. And, of course, this was only the beginning of the story. Architecture and, with it, the human appreciation of interior space continued to evolve.

This story of architectural-aesthetic progress leads me to ask, What about moral rules and systems? These too are products of culture, acts of the imagination. All societies have moral rules, but only a few have elaborated them into systems, into what might be called moral edifices. Are the people who live under large and complex edifices better off, more able to realize their full potential than people who live in structures of simpler design—moral lean-tos and huts? The answer is not at all clear. One reason is that large moral edifices are inevitably tied to sophisticated material culture. History is replete with examples of how the products of such culture, which include shrines, temples, churches, and mosques, can corrupt. Rather than inspire people to improve morally, they tempt them to vie for power and prestige. On the other hand, people whose moral edifices are as artless as their lean-tos and huts have been found to be gentle and caring, to value each other's company rather than material goods. Understandably, educated urbanites in both East and West have been tempted to romanticize them, and see only virtue in their lives.

But this picture does not bear close examination. Hunter-gatherers and other folks who live close to nature are human, after all. Certain behavioral traits, acceptable to them, are no longer acceptable to peoples elsewhere. An egregious example is the tendency of hunter-gatherers to be cruel to the deformed and handicapped in their own group and to be indifferent to the plight of all those outside their group. As for people who have been raised in elaborate moral edifices—in universal religions and philosophies such as Buddhism, Christianity, and Stoicism—they have glaring faults, which are well known, but they also have distinctive virtues, an outstanding example of which is their willingness to help the stranger. Evidence of this virtue in the landscape is the inns and hospices for needy travelers, the indigent, and the sick. The evidence goes beyond architecture, of course. As action, this virtue is most dramatically demonstrated in the way aid is

generously extended to victims of natural
disaster, even when they live at the other
end of the earth.

The Kernel

I concluded my lecture with the observation that the
three themes I had touched on were very different.
So the question arises, What do they have in common?
More generally, what do humanist themes have in common? My brief answer is that they all show a deep-seated
desire to understand the complexity and subtlety of
human experience, which in practice translates into
paying rather more attention to quality than quantity,
adjective than noun, psychology than economics.

Bilingual Presentation

The lecture was given in a way I had never tried before.
I read one paragraph from my English-language typescript,
and it was followed immediately by Liu Leng Xin's reading
of the Chinese translation. Would it work? I had my
doubts at first, for I thought that this double rendition
would slow the flow of thought too much for those who
already knew a little English. They would be bored.
But apparently not, from the feedback I had later. One
reason was the elegance of the translation and the fact
that it gave bilingual listeners the same message in a
slightly different flavor. Another reason was that the
three themes, though in themselves uncomplicated,
nevertheless were unfamiliar enough to bear repetition.
Moreover, I like to think that from a purely auditory

standpoint, our contrasting voices—male and female, one following the other—offered a not displeasing duet.

Any Questions?

The chair asked whether there were questions. From my experience in the United States, I didn't think there would be any. In a large, mixed audience of students and faculty, students tended not to want to risk making fools of themselves. So I was surprised when several hands were raised, and surprised even more by the nature of the questions, which dealt with ideas rather than, as was too often the case on American campuses, with simple points of fact. Moral issues were clearly important to my audience. One question was: "In a world of eating and being eaten, the foundation of survival for all animals, including humans, won't human morality soon reach a ceiling and what would that ceiling be?" My reply took on a Buddhist tone. I said something like this: "Because we eat, we should as moral beings allow ourselves to be eaten; indeed, offer ourselves to be eaten—to be severely drained of our energy to satisfy another's urgent need. All mothers, good teachers, and social workers know what I am talking about."

Flowers

When all was over, two bouquets of flowers arrived. After much picture taking with the bouquets as centerpieces, I handed them over, like royalty, to the students assigned to assist me. Copies of my speech in English were placed in a pile on the floor of the raised platform at the front

of the auditorium. Students rushed forward to obtain a
copy and then pressed around me for my autograph.
I signed the first few copies in Chinese and then reverted
to English to speed up the process, for I could see that
some students were growing a little impatient with the
slowness of my unpracticed Chinese hand.

At the Institute of Geographical Sciences

At 3:00 p.m., I went to the Institute of Geographical
Sciences and Natural Resources Research to give a
seminar at the invitation of Professor Li Xiu Bin, the
deputy director. A dozen faculty members gathered in
the conference room. I had hoped that I could use English
since these senior scholars were more likely than students
to know the language. But from my host's introduction, it
was clear that he expected me to do my best with Chinese.
The professors made it easier for me to get into the
groove by starting with factual, biographical questions.

Why had I switched from physical geography to
humanistic geography? I replied that I had always wanted
to do humanistic geography, but that I was temporarily
derailed by the beauty of the desert landscape. I fell in
love with emptiness and harshness—with death, or
rather, a mineral world in which there was neither
eating nor being eaten, neither decay nor cessation of
breath. That became a common state of mind for me in
adulthood. A premonition of it occurred, however, much
earlier, when I was a child of twelve or thirteen. In a
dream, I realized with preternatural clarity that being
alive could mean only one thing: that one day I would
die. Grown-ups sought to deflect my anxiety by saying

that I would outgrow it. Most children did as their curiosity turned from metaphysics to the opposite sex and to such solid school subjects as physics and chemistry. Mine didn't. I continued to be haunted by what I deemed to be fundamental questions. When the time came to choose a major in college, I suppose I could have chosen philosophy. But I didn't find academic philosophy's high abstraction, close argumentation, and radical questioning of experience attractive. So I sought to begin at the opposite end, with down-to-earth empiricism, and from there climb the philosophical heights. Although only sixteen, I knew that geography was the right earthbound subject for me. It would teach me the ways people struggled to survive against natural and socioeconomic constraints. But was survival all? What did people aspire to beyond it? What was their conception of the good life? How close was conception to reality? Questions like these would bring me to the foothills of philosophy and religion, where my ultimate interest lies.

The seminar continued for two hours and showed no sign of slackening. Li Xiu Bin, in his capacity as chair, said that we might take a half-hour break. I was so concerned to find the right words, or nearly the right words, for what I had to say that I forgot how exhausted I was until I struggled to stand up from the deep hollow of my chair and found that my wobbly legs could barely support me. I begged to be excused. The chair canceled the rest of the seminar and offered me his office to rest in. I stretched out on the padded sofa and fell briefly asleep. An hour or so later, refreshed, we all went to the Kong Yi Ji Restaurant for dinner. Kong Yi Ji is the name of a character in a book by Lu Xun (1881–1936), one of the foremost literary figures of modern China. Eating there was intended to give the occasion a flattering, literary aura.

A *Student Hands Me a Letter*

Back at the Jingshi Hotel, a young woman was waiting for me in the lobby. She handed me a neatly typed letter:

> Dear Prof. Tuan,
>
> Thank you for the lecture you gave us. I am really impressed by your insights of human [nature] and your way of perceiving the world around us. I heard from my supervisor Prof. Shi Peijun that you are planning to visit the Three-Gorge Dam. I wonder what you will feel about it, and what lead you to these feelings? As you know, there are already vast controversies about its impact on our society and environment. Would I be lucky enough to be the first person to read your thoughts on this huge human project? Wishing you a happy journey to the Three-Gorge Dam, and with my best regards,
> Jing Chen.

The Three Gorges Dam was on my schedule. I wondered myself how I would respond to it. With this specific student request, I realized that I couldn't just stretch out on a canvas chair on the deck of a riverboat, with my eyes half closed in luxurious indolence. I had to be alert! I had to think of an appropriate answer to Jing Chen's letter.

My Student Guides

Saturday was a regular workday in Beijing. But Sunday was a day of rest and recreation, and the State Laboratory for Resource and Environmental Information System dispatched a car and two students to take me to the Great Wall. I had already met the students, Zhi Cheng and Zuo Yi-Ou, for they were the principal translators of my lecture on humanistic geography. They were chosen then for their command of Chinese. Now, as my tour guides, they also had to have an intimate knowledge of the city and the ability to speak English. Both of my young guides were born and raised in Beijing and so knew the city well. As for speaking English, I wondered later whether they were relieved or disappointed to find that I could, after all, manage an inelegant but serviceable Chinese: relieved because they didn't have to use a foreign tongue all the time, or disappointed that they had lost an opportunity to practice it.

The Great Wall

There was a jarring clash of colors at the main entrance to the Great Wall: in the background were green hills and segments of the wall snaking up and down them; in the foreground were the gaudiest displays of commerce that I had thus far encountered. Merchants and artisans clamored for the attention of potential buyers, and they did so with every means at their disposal—big characters on banners and panels, loud primary colors, ceaseless shouting, waving arms and hands that reached out but did not quite touch the streaming tourists.

We pushed our way through the crowd, determined to ignore the hysterical sales pitches. We did stop, however, at an artisan's stall, fascinated by a man creating pictures of the Great Wall on slabs of black marble. He did so with a sharply pointed chisel that produced tiny dots. The work required keen eyes, a dextrous hand, and intense concentration. It was good to see someone totally focused on his art rather than on buyers. We stood behind him and looked on, captivated by the quick, machinelike, jabbing movements of the hand. We thought to ourselves that such work, hour after hour, day after day, with marble dust drifting into his eyes and breathed into his nostrils, would exact a heavy toll on his health. I felt I should buy something but hesitated because I didn't want to load myself with purchases so early in my China tour. At this point, he turned around and said that he would chip my name on it free of charge. I bargained in the local style and said I would buy one if he was willing to chip all three of our names on it. This he did. I observed happily to Zhi Cheng and Zuo Yi-Ou that henceforth our names would be linked forever by the Great Wall.

The climb up the steps of the wall was steep. I reached the third tower and admitted that I was short of breath and couldn't go farther. We stopped to look at the view—no, at history, for what made the view worth looking at was its history. I could almost hear horse hooves pounding the hills and Chinese commanders ordering soldiers to let their arrows fly. Both Zhi Cheng and Yi-Ou seemed rather excited by the visit. They had been here before, but long ago, when they were still young children. Zhi Cheng kept up a steady flow of historical information. I noticed that he spoke almost to the air, as if he didn't want to presume to be lecturing me.

My guides' manners were polite yet familiar—exquisite. I spoke Chinese with them. At times I clearly used a wrong word or idiom, which they chose to overlook when they understood my meaning. Occasionally, however, the mistake was so glaring and funny that they couldn't help laughing, as I did too when I recognized it. Here is one example. I asked them whether the Chinese still took their caged pet birds for walks in the park, as was customary. "Oh, yes," they said. "*Tsou niao*" (*Tsou* = walk; *niao* = bird). "What about dogs?" I asked. "Now that people are more affluent and have bigger living spaces, *tsou kou?*" That's when they laughed, for *tsou kou* (*tsou* = walk; *kou* = dog) meant something quite different: it was a term of abuse used against a servile person and did not at all mean "walking the dog."

Linguistically handicapped, physically I was at an even greater disadvantage, calling forth more examples of correction and consideration. I was a little unsteady walking down the uneven steps carved into the gangway between the wall's ramparts where it dipped too steeply. Zhi Cheng reminded me to use the handrail. He then took my other arm and guided me hundreds of steps to the bottom.

The Dog Doesn't Care Restaurant

For lunch, my young companions took me to a restaurant that specialized in meat buns, Chinese dumplings, as they are sometimes known in the West. The place bore the curious name "The Dog Doesn't Care." Of course, there had to be a story behind the name, and there was. The Chinese were and still are very fond of verbal games, using words not so much to explain as to add mystery to things. Couldn't McDonald's come up with fancier names for its eating places? I suppose it could, but that would not have served its purpose, which was to create an air of familiarity. Strange to think that America should strive for sameness and familiarity, and that China, contrary to its reputed desire for uniformity, should strive for individuality. Another thought crossed my mind. Humans are reluctant to acknowledge that they are animals, for animals eat, an activity that is seldom aesthetic. One way to repress the knowledge is cleanliness: food left partially eaten and plates stained with ketchup are quickly removed. This is America's solution. China's solution is, as we have seen, to give a literary flavor to restaurants. In extreme cases, the impression is created that one goes to them to compose poetry and drink wine rather than to engage in anything so crude as eating. Nevertheless, in elaborateness of aesthetic cover, the Chinese do not go nearly as far as the Japanese, who make eating into a quasi-religious ritual. Indeed, when I was confronted with "water-boiled fish" only four days before, I thought the Chinese did not go far enough for people of weak stomachs and overwrought sensibilities.

The cold dishes of preserved cabbage, cucumbers soaked in soy sauce, and thin slices of spiced beef stimulated my appetite. Then came the hot buns, biting into which squirted delicious juice into my mouth. Our driver ate quietly with us. He ordered garlic for himself. I remembered eating raw garlic with the meat buns when I was a child, so I asked for some—to Zhi Cheng's surprise. He probably didn't think I was Chinese enough to have a taste for it. So he peeled one for me. I could have done this myself, but to have a young man, seated to my left, do it for me was the sheerest luxury. When I had finished, Yi-Ou, the young woman seated to my right, handed me another wedge. Fed hand-to-mouth by youngsters, could I ask for more? Yet more was to come without my asking, as you will see.

Ming Tombs

The Ming tombs were on our afternoon schedule. The drive there would have taken us a half hour. Zhi Cheng used his cell phone to call A-Xing to report on our progress. Apparently A-Xing told him that I should be given an opportunity to rest. Up to this point of our tour, I sat in the front seat next to the driver so as to get a better view. That was Zhi Cheng's idea. Now he suggested that I change seats with Yi-Ou and sit in the back, where there was more room. I did as I was told, delighted to be so pampered. Then Zhi Cheng said, "You will feel more comfortable if you lean against me and rest your head on my shoulder." His startling considerateness sent a shock wave of happiness over me. Naturally I didn't— couldn't—accept even though I wanted to.

At the Ming tombs, we avoided walking through the richly decorated main gate. My guides let me know that we were entering the yin world of the dead and that it won't do to stride into it. Foreign tourists, who strode through the main gate, might have wondered why the Chinese so modestly sidled in. As we walked about inside the compounds, it occurred to me that my young friends weren't taking any pictures. At the Great Wall, they kept the camera clicking, but not here. I asked them why. Their answer was that we were in the world of the dead and it wouldn't be appropriate to record the presence of living persons there. Imagine showing one's friends a photo of oneself at the Tombs and saying, "Here I am, smiling among the dead!" I was impressed by the way Zhi Cheng and Yi-Ou, two bright university students, deferred to the traditions of their culture, including those that both communism and the scientific worldview have deemed superstitious.

Several flights of steps led down to the mausoleum of the Wanli Emperor (1572–1619), who spent eight million silver taels in a bid for immortality, enough at the time to feed a million people for six and a half years. Some thirty thousand laborers sweated six years to hollow out a subterranean palace for him and his two empresses. The result? I must say I was disappointed with what I could see—just three huge reddish-brown caskets. Even the treasures that accompanied the dead were not available for viewing; they too were hidden in reddish-brown caskets. As for the architecture, the arched ceilings of the five rooms of the palace were much admired. Arched ceilings implied a desire to create voluminous

interior space. In this aspiration, how successful were
the Chinese compared with the West?

Glass versus Porcelain Civilizations

My thoughts wandered to interior space, a topic I had
touched on earlier in my talk at Beijing Normal University.
When and where did the desire for it first arise? Could it
emerge in a civilization that promoted the cult of the
dead, the custom of burying important persons deep
underground—and, if not underground, then in the core
of a stone mountain, the pyramid? Egypt's pyramids
offered an exterior of sun-drenched splendor and, in
sharpest contrast, an interior of suffocating darkness.
Egypt's mortuary temples, too, contrasted a monumental
"outside" with a dark, cluttered "inside." Even Greek
architecture that no longer bowed to chthonian,
subterranean forces—even the Parthenon that rose up
to the blue Mediterranean sky—had dark, confined
interiors. The Romans were among the first in the West
to break away from these gloomy confinements to build
large, well-lit enclosed spaces, the bathhouses being
an outstanding example. But a truly sublime interior
that could lift the spirit to heaven was a creation of the
twelfth century, when glass technology enabled
Christianity not only to preach but to exemplify
architecturally a theology of light: sunshine filtering
through the rose windows of a great Gothic cathedral,
suffusing its vaulting interior with shimmering colors,
made it possible to believe that God was light and that
God created the world by driving away darkness.

China's civilization was a porcelain civilization, not one that rested on glass: a transparent medium that helped European science grow, as Alan Macfarlane and Gerry Martin argued in their book *Glass: A World History* (2002). By allowing light to enter and be a prominent element of interior space, glass had also given a certain slant to European pictorial art. Europe's theology of light, of which sparkling and transparent glass might be taken as a symbol, was absent from China, and that might be another reason why it was not driven to explore the aesthetics and symbolism of illumination.

These were some of my thoughts in Wanli's underground tomb. Would they have been different if I visited the Hall of Prayer for Good Harvest in the Temple of Heaven? This building had nothing to do with death and the underground. It was oriented to the sky, its symmetrical, hat-shaped roof being covered by sky-blue tiles. Inside it, looking up, I would have seen a domed ceiling supported by intricate carpentry, and I would certainly have admired it, but I would also have missed the play of light.

The Chinese famously saw yin and yang as equally necessary to cosmic order. In their daily lives, however, they couldn't help favoring yang. And so, at the Ming tombs, Zhi Cheng, Yi-Ou, and I sidled into the yin world of the dead, but when we were done there, we strode through the main gate back to the yang world of the living.

Playing Cards and a Jade Bracelet

A light rain rather than sunshine greeted us in the yang world, thus revealing once more reality's indifference to the requirements of symbolism. We sought shelter in the museum. Zhi Cheng bought a packet of playing cards in the museum's store that showed the faces of a long line of emperors and handed it to me. I don't remember thanking him. I seem to have arrived at a stage when gifts from him were taken for granted. He then dashed across the courtyard to a store, where he said he could exchange the admission tickets for various kinds of goods. Yi-Ou did not go with him. Instead, she opened her umbrella and held it over my head as we made our way back to the waiting car. She then went to look for Zhi Cheng. A little while later, both returned. Zhi Cheng had picked up a jade bracelet at the gift shop, which he gave me, saying it would bring me luck.

I was instantly reminded of the only other time I had had a jade bracelet. I was a six-month-old baby and wore one around each arm. My parents believed they would bring me luck. My arms grew so fat that one bracelet wouldn't come off even with the application of soap. The bracelet had to be broken. Did someone tell me the story? Surely I was much too young to register the event even if it had occurred. In any case, whether true or not, the story has become a part of me. I wondered from time to time whether breaking the bracelet to free my baby arm might bode ill for me in adult life. Well, it didn't, for my adult life has been remarkably free of major wrong turns and mishaps. Going back to China at an advanced age might have proved to be a wrong turn of my own choosing. But it wasn't so. And now I have another jade bracelet in my possession, given me this time by a Chinese student young enough to be my grandchild. Rather than seeing it as something that can bring me luck in the future, I see it as one more token of my good fortune up to this late point in life.

Lecture and Tours in Beijing

I was scheduled to give an informal talk to faculty and students at the Institute of Geographical Sciences. Any talk is "formal" to me, so I took the trouble of writing it up with the intention of reading from the prepared text. Almost at the last moment, I thought it a good idea to have the paper translated into Chinese so that the English and Chinese versions could be delivered alternately, as had happened at Beijing Normal University. Once more students were drafted to help. They came even though the request reached them without warning on a Sunday afternoon, when they no doubt had other things to do.

At 9:30 a.m., A-Xing and I were driven to the institute. We went into a seminar room. People started to pile in and soon filled it to overflowing. We moved to a large lecture hall. It too was quickly filled. The size of the audience made me glad that I had a prepared paper.

A Question in Human Geography

My topic is, What if, in human geography, the objects of our study are our intellectual peers? We take for granted that the scientist is superior to the object of his study.

"Superior" implies its opposite, "inferior." These words are highly charged. The use of them can offend, yet situations clearly exist where their use is noncontroversial. It is not controversial to say that humans are more intelligent than—and, in this sense, superior to—the inanimate world as well as plants and animals. We study *them*, not vice versa.

Young Children and Servants

Within the human world, I say confidently that parents are more intelligent and have more knowledge to draw on than their young children. This means that they can observe how their children act and come up with a theory—or, more modestly, a concept—concerning their behavior. In this sense, parents are the social scientists, their children the objects under study; in this sense, parents are superior to their young children.

In a traditional household, "superior" and "inferior" have a social meaning. A typical upper-class household in Edwardian England, for example, is divided into an upstairs of gentlefolk and a downstairs of servants. There can be no doubt as to who has the greater power, and with it, greater knowledge. But is it quite correct to say that the folks upstairs have the greater knowledge? For that notion is contradicted by the common saying that "no man is a hero to his valet." The saying implies that the master is no hero because the servant knows all about him—his strengths, but also his foibles and weaknesses. By contrast, the master knows little about his servants. And yet my claim still stands. For the master clearly has the means to study his servants if he so chooses. He can, for a start, make them answer questions, including highly personal ones. The servants, for their part, are not in a position to question their master the same way. From the master's viewpoint, knowledge about the servants just isn't worthwhile when he has the option to command.

Upper Class and Lower Class

Beyond the household, society as a whole may be strongly stratified. In such a society, the upper class can gather information, theorize and conceptualize about the classes below it, if it wishes. But the upper class almost never so wishes. Its reason, again, is "Why bother when we can command?" As for the subordinate classes, since they cannot command they observe. The more astute among them may try to map out their masters' routines, to understand and so anticipate their moods and behavior.

By necessity they turn into keen social observers, for their well-being depends on their ability to locate holes in the net of power that oppresses them.

Here we have a possible answer to the baffling question, namely, how is a successful revolution ever possible, for it is contrary to common sense that the weak can overthrow the strong. One answer is that the weak have allies among the strong. Another is the one I have just proposed: the weak may well enjoy the advantage of knowledge—critically, the knowledge of the human geography of their masters.

Superiority of Seeing over Doing

Notions of "superior" and "inferior" occur even in a society of sociopolitical equals. In ancient Greece, for example, philosophers may consider themselves superior to politicians and warriors, even though they are equal insofar as they are citizens. Philosophers see rather than do. Seeing—to Greek thinkers—is privileged over doing, because it supposedly leads to greater knowledge. Action, they say, is too passionately engaged, too confined to a narrow domain, to allow the sort of thinking and reflection that result in real understanding. A specific example is that of athletes competing on the playing field. They can know only in part.

By contrast, the spectator who stands aside and calmly looks down on the entire scene understands the whole. The word "theory" is closely related to the word "theater." The theorist is the spectator who looks down on the theater of life and tries to make sense of it.

The theater of life. That's the object of our critical and analytical gaze as social scientists and human geographers. But performances in the theater of life vary enormously in complexity. Students of the human scene, especially if they strive for clarity and comprehensiveness, are inclined to concentrate on the simpler ones. These fall under three categories: in the first are hunter-gatherers, pastoral nomads, and subsistence farmers; in the second are small entrepreneurs and businessmen; and in the third are the poor in stratified societies.

Observing Those Less Advanced

People in the first category were nonliterate and lived more or less isolated from civilization until well into the twentieth century. To study them, anthropologists and geographers borrowed a model, called ecological, that was designed by biologists for their work on plant and animal communities. Anthropologists and geographers knew, of course, that even people who lived close to nature in utmost simplicity possessed culture. But they didn't seem to think that the possession of culture in itself called for significant modification of the model. Indeed, culture was treated at times almost as biological extensions: the digging stick, an extra limb growing out of the cultivator rather than a tool made calculatingly to enhance food production; sapling-and-leaf shelters, "nests" in the sense that birds build nests, rather than architecture. As for stories, songs, and rituals—works that distinguish humans from other animals—older students of primitive life were inclined to see them, too, as adaptive strategies, the basic purpose of which was survival.

Scapulimancy as Randomization

The modern human geographer, in contrast to his forebears, doesn't see himself as a remote scientist looking down on the theater of life. He stays close to his subject; he is a participant observer. Naturally, he continues to observe, take notes, and reflect on what he has seen. But he may also do something from which scientists of an older generation refrain, namely, take the local people into his confidence so that, in time, the more intelligent and articulate among them become not just sources of information but fellow thinkers. Nevertheless, the modern social scientist—the anthropologist or human geographer—has to know more. That "more" is a distant and fruitful perspective not available to even the brightest local informant.

Let me give an example. The Naskapi Indians of Labrador practice scapulimancy. A shaman applies the burning point of a stick to a bone, producing cracks, which tell hunters which direction they should go on their next expedition. The Naskapi think that some benign spirit is guiding them. Now, a Western-trained scientist will sympathize with the Naskapi view, for he sees how it is part of a larger cultural outlook that sustains morale and livelihood. But he also sees something the Naskapi don't, namely, that scapulimancy is an instrument for randomization, the advantage of which is that it prevents hunters from going in one direction too often, thus exhausting the game (O. K. Moore, "Divination: A New Perspective," *American Anthropologist*, 1957, pp. 67–74).

Stability and Equilibrium

The more a phenomenon is repetitive, the easier it is for us to understand. Astronomy is our first genuine science, no doubt in part because the motion of the stars is, or can seem to be, undeviatingly repetitive. The phenomena of physical and biogeography are less orderly and predictable. Still, we discern enough order and repetition in them for us to think that we can study them scientifically. Understanding animal geography may be especially challenging, but it is made easier by the fact that orangutan social organization and the dam-building techniques of beavers have changed little over time. The human world is much less stable. Nevertheless, in earlier ages, change was slow, more oscillating than directional. Understandably, anthropologists and geographers sought to study these slow-changing communities and peoples through an ecological model in which a root concept was the balance of nature or equilibrium. The key words were adjustment, adaptation, and survival, and not, as might seem more appropriate when dealing with humans, planning, experimenting, and inventing. The native inhabitants themselves were more inclined to credit their culture to the genius of deities and heroes who lived in mythical times than to their own ingenuity. In doing so, they reinforced the bias toward stability or equilibrium that anthropologists and geographers brought with them from the disciplines of ecology and biogeography.

Shift in Conceptual Framework

A shift in conceptual framework occurred in the last third of the twentieth century. The root idea of equilibrium was found wanting. Flux, change, and nonequilibrium landscapes were introduced to a refurbished ecological science. These words and the concepts that went with them were brought to bear on the study of the human world, and not just the world of plants and animals that was its empirical base and original source of inspiration. But social scientists would have switched to a more dynamic model anyway, forced by what they see when they go out into the field. In the field, wherever that might be, it will be clear to them that small, isolated human bands that could have lived in near balance with nature no longer exist; and even if remnants still exist, by the late twentieth century, alien views and practices have penetrated them, disturbing significantly the "balance" they might have had. A consequence is that students of the human scene have no choice but to turn to larger, more complex groups— to agricultural communities.

The people in such communities can be quite sophisticated. By sophistication, I mean an awareness of a world beyond one's own, a habit of appraising that world, seeing opportunities there that can be used to improve one's way of life. Trade increases. A bus periodically rumbles through the street. Simple products of Western technology such as thermos flasks, bicycles, and radios are more and more common in the marketplace. More important than these material products is a way of thinking directed toward the future. Of course, all human beings anticipate events and make plans to meet them, hunter-gatherers no less than farmers and modern city people. The difference lies in the degree of the anticipation, in the scope and formality of the plans, but perhaps above all, in a psychological sea change such that individuals take pride in innovation, in having introduced new crops, found new markets, devised better means of cooperation all to further the cause of productivity.

Dynamic Societies

Now, suppose the scientist (say, a human geographer) turns away from agricultural communities to study dynamic modern societies, concentrating especially on their most innovative institutions such as the universities, the research centers, the great business corporations, or a whole region of creative vigor such as Silicon Valley. What intellectual advantage would he have over the people who work and live there? None that I can readily think of, if only because the researcher and the researched may well have attended the same universities and taken the same courses in engineering, economics, and geography. In the absence of a point on Olympus, the scientist is unable to offer a truly new way of looking at the world unknown to the locals. The best he can do is to present certain kinds of specialized knowledge and technical skills that will add speed and precision to the projects the locals undertake. Is this loss of status a cause for regret, or for rejoicing as we foresee a society in which there is true equality—intellectual equality?

Marxism and Maoism

Professor Li Xiu Bin, who chaired the session, asked for questions from the audience. Several were raised that prompted me to make an attempt at linking my speech more explicitly to Marxist and Maoist thought. This was easy enough to do, for Marxist geographers, more than other social thinkers, have emphasized asymmetries of power, as I have done in my own way. These asymmetries make exploitation of the weak possible. How can the asymmetries be reduced or overcome? One way is to empower the weak with an understanding of their situation, an understanding they don't have, but which an enlightened outsider may have and give. Karl Marx is such an outsider, and the enlightenment he offers is a theory of society known as dialectical materialism.

The point here is not whether Marxist theory is correct, but rather whether a new way of looking at society has been introduced. Marxist theory, genuinely original and widely accepted at one time, has had a huge impact on the world. Yet, apart from scale, it is basically the same story as the possible impact of the idea of randomization, if accepted, on the hunting practices and rituals of the Naskapi.

As for Maoism, someone asked me whether I am making the same point as Chairman Mao, which is that we can and should learn from the people. His question was prompted by my saying that in a modern society, scientists who study its most dynamic arenas will no longer enjoy a theoretical advantage over the men and women who live and work there; outside experts and well-educated locals become, in effect, coworkers. This attractive picture of equality pertains, however, only to the most technically developed parts of the world. Elsewhere, uneven wealth and intellectual resources are all too evident. We geographers have no trouble finding places that are backward compared to our own, and we continue to offer them not only practical advice but also fresh ways of thinking and analysis. "Teacher" and "taught" remain valid categories—an asymmetry in human relationship that we, in our passion for equality, strive to weaken.

Two Persistent Asymmetries

Two asymmetries will, however, always be with us. One derives from the fact that humans are animals, with strong biological drives and passions, not just minds. Although in the past three centuries our minds have embraced radically new ideas, our bodies and emotions remain relatively unchanged. The mind thus continues to see itself as superior to the body, able to rise above its exigencies and turmoils, and so study them in a cool and reasonably objective manner. Among the turmoils, the worst are violent conflicts that range from abuse and murder in the family to ethnic hatred and genocide. Since recurrences are easier to understand than novelty, it is just possible that one day we will truly understand their cause and cure.

The second permanent asymmetry is between ourselves and those who lived in the past. We dominate the past and the dead to a degree that we cannot dominate the present and the living. The dead are, in a sense, the ultimate hapless natives, observed but powerless to observe. True, with the lapse of time, we lose the inchoate details of life in earlier times, details available to those who were alive then. But we, their descendants, enjoy an advantage from that same lapse of time, which is distance and a broader perspective: we can weigh the significance of our ancestors, for we and only we can know the consequences (if any) of their existence.

The Sound of Turning Pages

As at Beijing Normal University, I read one paragraph in English and it was quickly followed by the Chinese translation. Turning the pages as I read, I was puzzled by the sound of page turning in the audience. What was going on there? I was told later that a mistake had been made. Students had access to photocopies of my lecture even before I spoke. They were following me line by line! If they already had the lecture in hand, why did they bother to come and hear it read? Just to see what I looked like? As it turned out, all was for the best. Students claimed that they benefited from the triple exposure. They had the feeling that they truly understood me. I was glad to hear this, for I have always aspired to be transparent in what I say and write, in compensation for the opacity of my lived life.

Peking Duck

For lunch, Dr. Zhou Chenhu, director of the State Key Laboratory at the institute, took us to a restaurant that was famed for its Peking duck. Imagine eating Peking duck in Beijing! In the search for authenticity, I needed to go no farther. Surrounded as I was by distinguished scholars at the round table, when the duck in its four glorious incarnations arrived (as crisp skin over a thin layer of fat, as tender meat wrapped in a thin pancake, as tender meat sandwiched in a hot bun, and as the principal ingredient in a soup), I felt completely justified in listening lackadaisically to the bright conversation so that I could pay greater mind to the food.

The Forbidden City

After lunch, the same two students—Zhi Cheng and Yi-Ou—volunteered to escort me to the Forbidden City. It was a hot and muggy afternoon. The two youngsters were concerned about how the heat might affect me,

since the city's vast open spaces had no trees and little
shade. Near the entrance, Zhi Cheng went to a stall and
bought a bottle of water. At the time I was mildly
surprised that, given his extraordinary thoughtfulness,
he didn't buy a bottle for me, too. How mistaken I was!
The bottle of water was for me. He just didn't want me
to have to carry it. He was my "water boy." From time to
time, he handed the bottle to me for a drink. He never
drank from it himself. When I asked him why, he replied
that he didn't want to yield to a physical need so easily.

The Forbidden City closed at five. We had only an
hour or so to spend there. I have seen photos, paintings,
and maps of the city so many times that the reality
before me made little new impression. I felt as though I
had simply switched from small-screen TV to panoramic
cinema. The scale of vision certainly expanded, but not
enough to make me want to sit up and marvel. The
reality of the Forbidden City also suffered by being
almost exclusively visual. Sounds and odors that would
have enhanced its realness were absorbed by the large
expanses of open space. Two features did stand out for
me in that hasty tour. One was the great, gold-plated
urns scattered in the palace grounds. Occupying Japanese
soldiers had scraped off most of the gold. One could
clearly see the scratches. The government, I assume,
left them there to remind visitors of Japan's greed and
China's humiliation. Zhi Cheng told me this story rather
than one from the scores he must have learned at school,
or could find in any guidebook. To a young patriotic
Chinese, the peccadilloes of emperors, eunuchs, and
concubines were remote events of the past, sounding
now like fairy tales, compared with the Japanese

occupation of China, though that too occurred long before he was born. The other feature that took me by surprise was the Imperial Garden at the back of the palaces. Its paths and partitions were conspicuously rectilinear and rectangular. Curves and circular shapes that one normally encountered in Chinese gardens were absent, as if to say that Confucian rectitude was imperative in the Imperial Garden and that round ponds and moon gates, inspired by Taoism, had no place in it.

Saying Good-bye

Back at the Jingshi Hotel, I went up to my room to rest before dinner. When at the agreed time I came down to the lobby, I found A-Xing and Yi-Ou, but no Zhi Cheng. Apparently he had something to do that night and had left. I was disappointed, for I had counted on having his company for a bit longer. In any case, I wanted an opportunity to thank him for his many kindnesses and to say a proper good-bye. I had his cell phone number, so I called him. We chatted awkwardly, and, finally, he ended the conversation with the traditional Chinese way of bidding a friend farewell, which is "May favorable winds speed your journey."

That was it. I felt let down, deprived of a satisfactory ending that could lend an extra edge to all that had gone before. I mentioned this desire for a proper ending earlier when I noted, with some distress, the inconclusiveness of the forum on urban planning at the Friendship Hotel. In real life, endings are seldom satisfactory—the curtain falls either too soon or too late, which, I suppose, is why I live so much in the rounded worlds of art.

Houhai Lake

That night, A-Xing, Yi-Ou, Qi Feng (a grad student at UW-Madison), and I went to Houhai Lake in the heart of Beijing, a popular night spot for young people in their twenties and thirties who earned a good income in the booming metropolis. The lake came to life at night, after about nine o'clock. Its shores were lined with beer gardens and coffee shops, all brightly illuminated by lanterns and strung-up lights. Boats of various sizes and shapes, some also brightly lit, plied the lake's dark surface, going nowhere. We hired a small boat and settled ourselves in cushioned chairs, between which was a small table. On it were bottles of fruit juice and soft drinks. A man at one end of the boat, using a long paddle, propelled us forward.

The night scene was unmistakably Chinese even though many elements in it were Western: the motorized cabs that competed with tricycles, the webs of colored lightbulbs that competed with lanterns, the beer and coffee, rather than rice wine and tea, that were being served. Even in our small Chinese-style boat, I had to drink prune juice, for there was no tea. On the other hand, the ambience of the place—crowds of people engaged in quiet entertainment outdoors, well after dark on a workday (Monday)—was not a scene one would find in Western capitals.

Good-bye Beijing, Hello Chongqing

Wang Fu Jing Shopping Plaza

We were to depart for Chongqing late in the afternoon.
As the morning was still free, A-Xing suggested that we
visit a popular shopping plaza next to the old Foreign
Legation Quarter. The name of the plaza? Wang Fu Jing.
There is, as with so many Chinese places, a story behind
it. I am not sure who Wang Fu was, probably a successful
entrepreneur. In any case, a Wang Fu Street existed in
the Yuan dynasty (1279–1368). When a well *(jing)* was
discovered there in the Ming dynasty, the name was
changed to Wang Fu Jing. By the late Qing dynasty, the
street was a busy commercial thoroughfare, lined with
shops and goods that became brand names in China.

Wang Fu Jing boasted a large bookstore. Naturally we
went there, but we did not stay long. I, for one, didn't
want to wander among the counters for fear of finding
The Da Vinci Code or the latest Harry Potter on display.
An old prejudice of mine reared its head. Foreign cars
and foreign stores (including Starbucks and McDonald's)
were quite acceptable to me. But for foreign books,
especially foreign best-sellers, to be given pride of place
in the bookshop of an ancient literary civilization? No
way. My mind drifted irrationally to the unhappiness
many Chinese felt when the Nobel Prize was awarded
to the American author Pearl Buck in 1938 for her

novel about Chinese peasant life, *The Good Earth* (1931),
as though no Chinese writer then alive was worthy of
the honor.

We dived into a basement food court for lunch.
Chinese foods of all sorts were being cooked or were on
display in the stalls that lined the four sides of the
cavernous room. What made the place feel Chinese was
not just the food but the pungent odors, as well as the
swarms of customers moving around to find a seat, carrying
with them—precariously—their bowls of beef noodle
soup and small dishes of chicken feet, sautéed shrimp,
cabbage rolls, sea cucumber, or whatever it was that took
their fancy. I ate, wedged in between Yi-Ou and A-Xing.
I was glad I didn't have to use the toilet. Even if there
was one, I wouldn't have wanted to imagine its condition.

Back at the hotel, I said good-bye to Yi-Ou by
presenting her with a copy of my Princeton lecture, now
published as a little red book called *Place, Art, and Self*
(2004). I very much wanted to give it to her and another
to Zhi Cheng, but I had only one copy with me. I solved
the dilemma with an inspired inscription.

Bidding Farewell to Two Students

To Yi-Ou and Zhi Cheng:

Just as our three names were linked by the Great
Wall [I was referring to the carved black marble],
so they are now linked in this little book.

Fondly, Yi-Fu

Although I knew these two students for only two days,
their impact on me has been profound. In the two

months since I have been back in Madison, I think of them almost every day. Their images, however, are fading, hard as I try to retain them. My hope now rests on the photos that were taken of us. I look forward to sticking them on the fridge.

How did people remember each other before photography was invented? Chinese poems are full of sad partings. Strange to think that in premodern China, when a dear friend, after saying a last good-bye, turned around to climb on his horse and ride into the distance, the details of his face would immediately recede, never to be recovered until the next meeting, which might have been years in the future.

Meeting Ouya, Samuel, and Alex

That afternoon, A-Xing helped me check out of my hotel. Together we went to his Beijing apartment. There I found Ouya and the two children, Samuel and Alex, who had arrived from Madison on the evening of June 3. The Zhu family is remarkable in several ways, one of which is that its members are citizens of three different countries. The parents are Chinese; Samuel, born in Toronto, is Canadian; Alex, born in Madison, is American. They are in themselves a perfect example of international cooperation and harmony. I carry an American passport, so I sometimes tease eight-year-old Alex by saying, "We Americans must stick together." At customs, we show immigration officers passports that, for all their differences in color and language, say some variant of "the Secretary of State [the Secretary of External Affairs, or the Minister of Foreign Affairs] requests that you, the foreign government, give assistance to our citizen in case of need." What if the foreign government doesn't oblige? Well, these polite requests must ultimately be backed by force. In a fanciful mood, I see A-Xing and Ouya as backed by the People's Liberation Army; Samuel, in a crunch, can appeal to Her Majesty the Queen; Alex and I, for our part, have the clout of the Seventh Fleet.

Chongqing: The Shock of Change

The biggest material change I noticed in China was the airport in Chongqing. Was there even a civilian airport in 1941, the year I left China with my family? I don't remember any such thing. I do remember hurrying to the airfield in the dark. We arrived at the edge of an airstrip and waited outdoors with our battered suitcases, along with a few other passengers. Could this picture be correct? I say "outdoors" because the sweet smell emanating from the rice fields and the sound of croaking frogs were part of my memory. We waited nervously for the sound of an approaching airplane and scanned the black sky for lights that might spell our salvation, but saw only the stars. Two stars, however, appeared to be moving. They moved closer and closer until the faint outline of an airplane could be seen.

Why the nervousness? Why the hush that parents imposed on their children? We were listening for Japanese airplanes that ruled the sky over China's wartime capital. Our ears were attuned to the sound of the air-raid siren, which came on quietly and then swelled to a scream, and, less certainly, to hear or hallucinate a distant, low-pitched drone that might signify the coming of a flying armada. These sounds haunted our everyday life. If we heard them that night in 1941, it would have meant the collapse of our hope for escape, and no one knew when the next opportunity would come.

In 2005, all that seemed a bad dream. For, of course, our plane did not land in a rice field. Rather it touched down on an expanse of concrete, with crisscrossing runways illuminated by strings of orange lights; and as the airplane taxied to the gate, the flight attendant delivered the reassuringly familiar minispeech to the effect that passengers should remain in their seats until the plane had come to a complete stop and the captain had turned off the seat-belt sign. I could have been in Chicago, except that the speech was delivered in Chinese.

A minibus from Chongqing Normal University awaited us at the airport. It took us through islands of bright neon lights alternating with patches of a countryside steeped in impenetrable darkness. Greater Chongqing was a cluster of cities, lodged on the slopes and floors of mountains and valleys. At the center, where the Jialing and the Yangtze rivers met, was the old core. What did Chongqing mean to me as a child? We lived in the suburb, so I was seldom in the city itself. Nevertheless I retain memories of thick humanity in a maze of shops, steeply inclined streets, steps everywhere, people being carried in sedan chairs, and signs that directed people to air-raid shelters. Incongruously, I also retain an image of glamour—the effect of one surreal night when our parents took us into the city to see *Snow White and the Seven Dwarfs*. The movie was sponsored by the British embassy in support of a charity for wartime orphans. At the conclusion, the wife of the British ambassador appeared on stage to make a little speech. I thought she was Snow White herself—she was so beautiful in her shimmering blue gown.

That was 1940. Sixty-five years later, I was back in Chongqing, rubbing my eyes to wake up as the minibus stopped at the door of the Harbor Plaza Hotel. Alex got out and strode confidently past white-gloved doormen to the revolving door. I caught up with him and turned around to say to his parents, "Our Alex knows where he belongs—a five-star luxury hotel!"

The Zhu Children

"I hope you don't mind traveling with the children," A-Xing said to me in Madison. "Not at all," I replied politely, not realizing that politeness could have its own wisdom, for it turned out that Samuel and Alex, amply endowed with that innocence and goodness peculiar to childhood, were able to impart a rosier hue to my own outlook on life. Of the two children, the younger one, Alex, is very shy and quiet. He has the sweetest smile, which he bestows on his favorite people with a slight downward tilt of the head. He speaks so softly—if at all—that to hear him you have to bend down to his level. Yet even in unfamiliar places he moves through the crowds with assurance. The backpack he carries (I have never discovered what's in it) also gives him the air of an intrepid explorer. Alex must have concluded that social space requires more circumspection, a finer antenna, to negotiate in than does physical space.

The older child, Samuel, is quite different. At thirteen,

he is as tall as his father, strong, and his voice is
deepening. The dark blue sweater he wore on the first
day of the trip read, "Future Engineer of Wisconsin."
He won it in a mathematics competition at his middle
school. Samuel is the ideal big brother, for he is not only
protective and caring but also happy to spend time with
his shy sibling. I often see them together, sitting a little
apart from others, playing a sort of game, which may be
no more than Samuel pretending to punch Alex on the
arm and chest, making him squirm and smile.

In his relationship with me, Samuel was all courtesy
and care. He stood aside to let me through the door, he
warned me of roadside curbs and slippery patches on the
sidewalk. He might carry a Canadian passport, but he
behaved very much like a child brought up on the ideals
of traditional China. Certain of his actions were, however,
so spontaneous and delicately attuned to a particular
situation that they couldn't have been taught. For
instance, at the front desk of our hotel a bowl of brightly
colored candies invited guests to sample while they
checked in or out. Samuel took one, and I followed his
example. I wrestled with the cellophane wrapper. He
immediately volunteered help. At dinner, when I
expressed an interest in coffee, he offered to go to the
counter and get me one, making sure before he took off
whether I wanted sugar and cream with it. On his own
initiative, he brought me an Oreo.

At one point on our trip, A-Xing loaded heavy
suitcases and bags onto a cart that the airport provided.
After one of my praises for Samuel, which I sincerely
meant, A-Xing (as the proud but modest father) pointed
out that his son didn't help him with loading the baggage.

I was baffled for a moment, then the obvious explanation came to me. All his life, Samuel had observed his father doing the heavy work. Seeing his father loading suitcases was therefore not a behavior that would catch his attention. What could and did catch his attention, because it was not a daily occurrence, was his father's solicitousness toward me. Helping a stranger might just be in Samuel's nature, but the desire would atrophy if it were not reinforced by example from someone he admired.

Touring "Authentic" Chongqing

Breakfast in a Food Stall

Rather than have a fancy breakfast in our hotel, which was what we did in Beijing, A-Xing called us to adventure. He suggested that we (that is, he, his family, and I) go out and try one of the food stalls that lined the busy streets. As soon as we stepped out of the cool fragrance of the hotel lobby, we plunged into a world of humid heat and sour odors, the milieu of the poor impinging on the milieu of the rich so characteristic of cities in developing countries. We soon found a food stall, a modest establishment of four tables. We could see what it had to offer, for next to the tables were a man and a woman, one stirring porridge in a huge black pot and the other plunging dough sticks into boiling oil.

A distant analogue of the Chinese food stall is the American diner, which I patronized regularly in the 1950s. When the matronly waitress behind the counter came by with a coffee pot and tipped its spout into my slightly cracked cup, she filled not only my cup but my sense of well-being to the brim. The diner had a distinctive odor—a mixture of fried eggs, bacon, hash browns, and (in the old days) cigarette smoke. Absent were the rancid notes of poverty. Moreover, despite the cooking odors, despite the hand-me-down furniture and yellowing framed photographs of local sports heroes, I never questioned the diner's hygiene. In fact, I suspected that its warmed-up Campbell's soup was more wholesome than a fancy restaurant's bouillabaisse, into which a vengeful waiter might have spat.

My confidence in Chinese food stalls was less than full, however; and it didn't help when A-Xing stopped me from laying my chopsticks on the table, saying that it might not be clean, and that it was advisable to rest their upper ends on the edge of the plate.

Chiang Kai-shek's Wartime Residence

That morning our plan was to visit Huang Shan in a secluded part of Chongqing, where Chiang Kai-shek held court in China's war with Japan. Our minibus climbed a mountain and stopped at a courtyard in front of a gateway. This was as far as it could go. Beyond it lay a footpath and many steps. It's a good bet that the Generalissimo and his glamorous wife, Soong Mei-ling, never left their footprints on them. Chinese officials were borne up in sedan chairs. Did General George Marshall also travel on human shoulders? He was in Chongqing to persuade the Nationalists and the Communists to cooperate in the war against Japan. His house stood next to the Generalissimo's in a secluded wood. Chiang's house had to accommodate his assistants, personal guards, and servants. Given that necessity, it was surprisingly modest. Workmen swarmed all over the place when we were there, sprucing it up for tourists, especially those who were expected to come to China for the 2008 Olympics. At least, that was what I heard. I wondered who the tourists would be—surely mostly

people more or less my age who had personal experience
of the war with Japan? But by 2008, they would be far
too old to climb the steps. They will have to be carried
up in sedan chairs, thereby making their experience
unwittingly authentic.

On our way down Huang Shan, A-Xing pointed out
to me several modern yet modest-looking houses. These,
he said, claimed to be farmhouses but weren't. They were
rather retreats built for rich city people tired of glitz
and life in the fast lane. Was this, I wondered, another
example of how quickly the newly affluent entrepreneurs
picked up Western ways, or was it a reversion to the
Chinese tradition of withdrawing to the farm, becoming
a Taoist, when life as a Confucian scholar-official at
court became too burdensome and even dangerous?

Authentic Sichuan Food

We decided to have authentic Sichuan food for lunch.
Our driver, a local man, took us to a place that catered
to well-off locals rather than tourists. We were led to a
private room—one of a succession that lined both sides
of a long corridor. Dining in private appears to be the
norm in better-class Chinese restaurants, and I have
wondered why this isn't the custom in the West. Is it
because eating in China is an informal affair, with diners
sucking noisily on chicken feet and spewing out shrimp
shells, that is best done, as with all things biological,
behind closed doors? By contrast, dining in the West has
evolved into a sort of public ceremony, perhaps because
the earliest restaurant chefs, before the French Revolution,
served in royal and aristocratic households that made a

point of refined etiquette. Eating was a performance: deftly
separating fish bones from the meat with a fish knife
merited, as it were, an appreciative audience.

In a typical Chinese restaurant, a round table occupies
the center of the private room. A large circular plate
that can be rotated—a modern invention—takes up the
center of the table. As each dish comes, the host rotates
it to stop before the guest of honor, who is obliged to take
something from it. In time, the entire plate is covered
with food. A delicate point for diners to consider is how
to turn the plate so that one's favorite dish stops before
oneself without placing the less desirable offering before
another guest. Subtle competition for the favorite—and
there is bound to be only one or two such—makes
politeness almost a moral challenge. It is much easier for
the Western diner to be polite. Since food is served in
individual portions, he can safely engage a neighbor in
conversation with no thought as to the predatory designs
of other guests.

In the interest of authenticity, we asked our
Sichuanese driver to order for us. When the dishes
arrived one after another, I found I could hardly eat
any of them, for all seared the tongue and burned the
throat. I was also discouraged by the size of the servings,
each a bulging mound, whose sauces oozed to and
beyond the edge of the plate. Finally, I was repelled by
the local delicacy—fried eels, which had been blackened
by being fried in a dark spicy sauce and came in coils,
with heads and tails intact.

At our table were A-Xing, Ouya, Samuel and Alex,
the driver of the minibus, a lecturer from the local
university who was our host, and me. Only A-Xing and

the driver could do justice to the gargantuan meal. My own efforts were puny. I was saved, however, from a total loss of face by Alex, who showed interest only in fried rice and Sprite. At the end of the meal, the mounds of food retained their topography, with only a dent here and there to remind one that some sort of halfhearted mining had occurred. Fortunately for our consciences, the driver was more than happy to take the food home.

That evening, upon my pleading, the Zhu family and I went to a coffee shop near our hotel for something to eat. I couldn't stand more authenticity—more coiled eels! I ordered a slice of chocolate cake to go with my coffee and shared the cake with Samuel.

Revisiting My Childhood: Nankai Middle School

At ten the next morning, A-Xing and I had an appointment with Song Po, the principal of Nankai Middle School. When we met, I thought he was about the age of my older graduate students. Increasingly, even the most responsible members of society—professors, deans, directors, and principals—looked preternaturally young. One advantage of being old is that I am now less self-conscious and socially diffident. We sat down to tea in the reception room. A brief ceremony of gift exchange followed: the principal gave me a book that celebrated the one hundredth anniversary of the school, and I presented him with a framed picture of three distinguished alumni of Nankai, taken in 1940 against the backdrop of Jin Nan Tsun, the school's dormitory for teachers. The alumni were the mayor of Chongqing, the Communist leader Chou En-lai, and my father, a Nationalist official. There they were, with their spouses, a picture of old-school-tie comradeship that momentarily transcended their opposed political ideologies.

The principal asked whether I would be willing to speak to the students. Giving a speech was not in the program, but I agreed to do it, in part from politeness, in part from vanity—the desire to make an impression on schoolchildren and not just on scholars. For my topic

I offered "Coming Home—after Sixty-four Years." Song Po said that he would let the school know through the public announcement system, and he thought it best if my talk were the last event of the visit, after the campus tour and luncheon with him and his staff.

The Campus

Seeing the campus, the last place in China I could personally identify with, was, of course, the main purpose of the visit. I wanted to wallow in nostalgia, let memories bubble up from the deep well of my being, hold hands with my childhood self. But these desires were not easily accomplished, if only because the teacher assigned to show us the campus was anxious to have us admire the new facilities—the bright science labs, the well-stocked library and museum. As we walked about, certain familiar features did cross my field of vision. One was the courtyard behind the school's main entrance. There the bus used to stop every Friday, bringing Father home from work at the Foreign Ministry. Another was the large oval playing field, which still dominated the campus. In the midmorning hour that we walked by, it was empty, which allowed me to invoke the image of a scrawny child who sprinted along one of the tracks when no one was looking, pretending to be the school's champion. I winced at my young self's total lack of realism. Next to the playing field was a fishpond that I had angled in illegally with my younger brother. We dug earthworms from the moist soil and used them as bait. At that age, I thought nothing of tearing an earthworm apart and sticking a still wriggling half on the sharp end of a bent pin. I have too much imagination to do that now. Ignorance of the pervasiveness of pain in the world must lie behind children's innocence, their capacity for happiness. Next to the pond was a small, deeply shaded woodland. Did I dream of elves and fairy romance there? No, I didn't. Nor did my friends. We were boys and liked to play rough. We went to the woods to look for V-shaped branches that could be broken off and, with rubber bands, made into slingshots. Our victims were mostly birds and other small animals, but in the heat of a war game we didn't hesitate to zing pebbles at each other.

Ghosts of the Past: Dr. Chang, My Father, and Chou En-lai

My sharpest memories are of Jin Nan Tsun (*tsun* = village). As I have said, it was a dormitory for teachers, but it also served as a residence for friends and protégés of the school's commanding founder, Dr. Chang Po-ling. My father attended the original Nankai School, the one that was built in Tianjin in 1905. Dr. Chang became Father's patron and then simply his friend. That was why when the government moved to Chongqing in 1938, my father and his family were assigned a small courtyard house in the dormitory of the new Nankai School. Our house was number 7 in the second row, right behind number 1 in the first row, where the headmaster, Dr. Chang, lived.

Dr. Chang was the only one in the village who had a refrigerator. On a steamy summer day, he might generously send a tray of ice cubes to his neighbor. One such day, a tray came to our house, but I was not at home. I had gone out to watch a basketball game. Father desperately tried to save a cube for me. He thought the ice would melt more slowly if it were put in salt water. When I finally came home, all that was left of this elixir from heaven was a sliver of ice floating in a glass of salt water.

In the past twenty years, old school buildings had been torn down and new ones put up. But old and tattered Jin Nan Tsun survived. It survived—it is being actively preserved—for historic reasons. Important personages, both Nationalist and Communist, had lived there or visited it, the reason being the charismatic presence of Dr. Chang. Chiang Kai-shek came to Nankai School to pay his respects. Chou En-lai, Father's schoolmate in the old Nankai and Mao Tse-tung's front man in the Chongqing negotiations, came to visit frequently. He was often in number 7. Known for his charm throughout the world when he became China's premier, he charmed my siblings and me in the years 1938–40 with toys and stories. We called him "Uncle Chou," as was the Chinese custom. Strange are the things we remember. My brothers and I, now in our seventies, recall the occasion when Chou En-lai and Father arm-wrestled at a small table in our humble home. My father wanted to test the strength of Chou's hurt but supposedly mended arm. The symbolism of this small event probably helped to fix it in our memories. At this time, both Chiang Kai-shek and Mao Tse-tung were in Chongqing, trying to establish an alliance against their common enemy, the Japanese. A pity they couldn't use arm wrestling to resolve their differences.

Our Old Home

So how did I feel when I reentered Jin Nan Tsun and stood at the door of my old home? I felt simultaneously and contradictorily both more unreal and more real: I was a ghost trying to relive the world of a child and an old man given a fresh burst of life by infusions of odors and tactile sensations from the past.

Visually, our house looked so small. I know this is a common response of people who revisit the places they lived in as children. But it is not true, for me, in all cases—not true, for example, of the playing field, which still loomed large. As for the house, how could eight people—our parents, four children, and two servants—fit into that small space?

We pushed open the door of number 7 and entered. People lived there, though we didn't meet anyone that day. I lived there! The little courtyard was completely paved over. In my time, it had two patches of exposed earth next to the outer wall. There my younger brother and I tried to grow watermelon, our favorite summer fruit. We watched over it every day, impatient for it to swell into edible size.

The courtyard looked cluttered, as were the dark rooms that surrounded it on three sides. How come the rooms were so dark? My few memories of what took place in them were either daylight scenes or night scenes, but well lit. They were an odd assortment, like the knickknacks one might find in a neglected desk drawer, and included my siblings and me being bathed one by one in a wooden tub, sleeping with a can of chirping crickets under my pillow, painting watercolor landscapes with the brush and paint box that my parents bought me for my tenth birthday.

Besides the old home, I wanted to revisit at least two other places—Nankai Elementary School and the tree-lined road that led from our home in Jin Nan Tsun to Nankai Middle School's main gate. Elementary school? At this point, I can no longer postpone confessing that I never was a student at Nankai Middle School. The principal would have known had he consulted the records. In 1941, I was ten years old, still too young and unprepared to go to Nankai. My elder brother, older than me by one year and three months, did go; and in 1997, when he visited the school he found in the records his name and the grades he earned. The elementary school was founded by parents who wanted the prestige of Nankai but whose children were too young to attend. Dr. Chang allowed an informal association between the two institutions, and so, in that sense, I was a Nankai boy.

We followed the path to the elementary school. It was still much as I remembered it, as was the hamlet— Pei Shu Tsun—at midway. The path was blocked, however, by a row of new houses and by a busy street. In any case, there was no point in going farther, since the one-room school by the electricity-generating station—my school—had been demolished and replaced with a completely new building.

As for the road that led from our home to the main gate and ran through the school grounds, we caught glimpses of it on our tour, but we didn't have time to follow it. The principal's luncheon, our guide reminded us again, beckoned. This might seem a cause for regret. On the other hand, had I followed the road, the worst nightmare in my entire life would likely have returned to haunt me.

A Nightmare Recalled

In the dream, I walked down the road to meet Father's bus as I had always done with my brothers. But this time I was alone. It could be that I cunningly excluded my brothers so that I would be the only son there. As I walked, the sky darkened, a thick fog rolled in, and the air felt increasingly chilly. I walked faster, growing anxious and scared, until I saw Father in the distance. The bus must have arrived early. I ran toward him in relief. When I got close I saw to my horror that it was not Father—it was a figure whose feet did not quite touch the ground and whose face, though it had Father's features, could barely be recognized because it was in an advanced stage of decay.

I was nine years old at the time. I woke up, my underwear drenched in sweat. I forced my eyelids open for the remainder of the night. The next morning, I felt better, yet unlike past bouts with bad dreams, dread lingered even when I was at school and attended to the normal affairs of the day. As night approached, I became fearful once more. A cousin (Chang Shouyi), several years older than the Tuan siblings, was visiting. We loved her visits, both for her delightful self and for the stories she told. She told them sitting on a stool placed either in the courtyard or just outside the gate, with the number 7 on it, to catch the breeze. That evening, when I found her alone, I told her my nightmare. In the course of telling, I became increasingly agitated and started to cry.

The walks to the main gate to await Father's return from work were happy events. My brothers and I looked forward to them. We walked, ran, and chatted, as children are wont to do. We often had to wait, but when the bus finally rolled in, we rejoiced. We watched the passengers filing out of the bus one by one. They all looked unreal until Father himself emerged. Why did a mere bad dream have the power to cancel reality? Why couldn't Father's numerous kindnesses to me, including his heroic effort to save an ice cube, erase this horror from my memory? Even now, the nightmare can return with such vividness that I feel goose pimples rising on my skin.

Lunch with the Faculty

Luncheon took place in a room next to the students' dining hall. At about noon, boys and girls straggled in. Since they were all dressed like American teenagers, I expected them to eat pizza and submarine sandwiches, but, no, the power of fashion did not extend to food. Chinese youngsters, for all their backpacks and ponytails, stuck to the dishes their mothers cooked: Sichuan cabbage and shredded pork, smoked fish, beef noodles, wonton soup, meat dumplings, and other typical Chinese fare. At our round table sat the principal, Song Po, three teachers, a dean (Li Ching) from Chongqing Normal University, A-Xing, and me. I had all sorts of questions that I wanted to ask about Nankai School—how it was financed, where the students came from, what the school fees were, how the curriculum was determined, who decided on what textbooks to use, how important

competitive sports and extracurricular activities, such
as drama and music, were. It would have strained my
Chinese not only to ask these questions but also to
understand the responses fully, for these necessarily
included technical words and expressions that I did not
know. Throughout the trip, I had to bear with frustration
of this kind. My eagerness for information always far
exceeded my language skills to inquire and to receive.

At lunch, I was also conversationally hobbled by my
need to rehearse what to say to students that would
engage them for half an hour. No one seemed to know
how many would attend, but I was told that they would
be fourteen- to sixteen-year-olds, for the older ones had
just taken their nationally administered exams and had
already left for the summer break. So I was to have restless
midteens for my audience! That in itself seemed daunting.
On our way to the classroom, the geography teacher
expressed the hope that I would say something about the
field of geography. I asked him how geography was
taught. The answer he gave made me think that it was
taught in a standard way; that is, there were lessons
on climate, landforms, soils, populations, economic
activities, and so on. I moaned inwardly, for how could
I possibly make these weighty topics interesting in the
short time I had, and, moreover, how could I make them
fit into the framework I had come up with?

Speaking to Students

I entered a large classroom and could see at once that it
was packed with students—possibly a hundred of them—
all bright eyed and bushy tailed. I use this familiar idiom

because I can't think of a better way to describe the atmosphere. All public speakers know what I am talking about—how one can immediately sense the mood of an audience, whether one has to struggle to overcome its resistance, or can count on being buoyed by its goodwill. I sensed the goodwill, and I now wonder why it was there. It could be that the youngsters welcomed a break from school's routine, or wondered what an aged American-Chinese would have to say about their home, their world, for the talk's title contained the words "Coming Home." It could also be that they were curious about a man—obviously a Chinese—who nevertheless would give his talk in English, and they, being smart and ambitious, wanted to put their knowledge of English to the test.

While we toured the campus and also while we had our lunch, I decided that my theme would be "home" and "world"—or rather, the move from "home" to "world." With home, I could establish what I had in common with the students. Reminding them of that commonality right away was essential. I told my young listeners that their school, their home, was also my school, my home, that between the ages of seven and ten, I lived with my family in Jin Nan Tsun. I told them about growing watermelon in our house's small courtyard, fishing in the pond next to the playing field, and going to Nankai Elementary School, following a path that led through Pei Shu Tsun. All these places the students also knew well. But, then, I said, I left home for Australia, the Philippines, England, and the United States. As I grew older, my world opened up. I saw things and learned about things that I had not dreamed of.

The move from home to world proved to be enormously enriching.

At this point, I addressed my audience directly and said something like this: "I wish your life would follow a similar path, that after graduation, you would take on the world. And by this, I don't just mean Beijing and Shanghai. Of course, you are drawn to these great metropolises, and you certainly ought to go there. But while you are young and on the move, why not stretch your imagination? Why not go farther afield to places such as Tibet and Inner Mongolia? Study these places, learn something about their climate, landforms, and ways of livelihood. With such knowledge, you will not just be a tourist having a good time, you can also be of service to society."

(As you the reader can see, I have slipped in a bit of standard geography here to please the geography teacher. At the same time, I was able to increase my hold on the students' attention by making an appeal to their idealism, their patriotism, their desire to be of service.)

Being out in the world, I continued, has its downside: "One is always an outsider—an observer—looking in. One is never a native, someone who knows the human richness and intimacy of a place and its people from the inside. For all that I have learned in the world and about the world, at age seventy-four, I begin to wonder whether I have become rather superficial. I don't want to die feeling superficial. That's why I have returned home. That's why I am here with you today."

With that, my talk ended. "Are there any questions?" the chair asked. I didn't think there would be any. How could there be? These were Chinese children. I addressed them in English, which meant that their questions to me would also have to be in English. To my surprise, hands started to wave. A boy asked, "Well, what did you find when you returned home?" The question took me by surprise. I had ended my talk with a rhetorical flourish, hoping that my audience would rest content and not want to probe further. But no. Fortunately, two words came out of the blue to my rescue. They were "anchorage" and "tenderness." That's what I found when I came back to China, I replied. Another student, a girl, wanted to know how I coped with racial prejudice. I probably disappointed her when I said that I rarely encountered it in person, adding that even if a racial slur were directed at me, I probably wouldn't recognize it, so full was I of myself as the inheritor of a glorious civilization. The final question was a fluff ball: "If I want to study geography in America, which university should I go to?"

In a way, I shouldn't have been surprised by the students' alertness and curiosity. Nankai has always been a good school. It was so when my elder brother attended it in 1940. It still is. Every year it sends thirty to forty students to Bei Da and Tsinghua, the Harvard and MIT of China. After the talk, students gathered around me for more exchange of views. I wish I could have lingered with them, for nothing stirred me to life more than the eager, probing questions of the young. I went back to Nankai for a taste of the past. I got that, but what I didn't expect was the bonus—a taste of the future.

Preparing for the Riverboat Trip

Back at our hotel, we packed in preparation for the next stage of our journey, which was a boat ride on the

Yangtze River. The logistics of the move to the riverboat were rather complicated. What happened was something like this, and I am giving the account not for its inherent interest (it didn't have that) but rather for showing once more how utterly spoiled I was on my trip to China. A-Xing and Samuel hired a cab. They put all the bulky luggage into it and had the cab take them to the dock. There they checked into the riverboat and made sure that the two cabins, one for the Zhu family and the other for me, were in good order. Meanwhile, Ouya, Alex, and I went into a coffee shop, where we waited in air-conditioned ease for A-Xing and Samuel to return and pick us up.

The two days in Chongqing and, above all, the six hours in Nankai Middle School were the culminating points of my journey into the past. It was there that I spent my final three years in China before leaving for foreign countries. There, too, I had stayed long enough and was old enough to have accumulated lasting impressions. The next two stages of my journey—the riverboat ride from Chongqing to Yichang and the flight from Yichang to Shanghai—would be new experiences for me. How would I find them? Would I respond to the Yangtze River and the Three Gorges as any tourist would, or would there be a faint cast of nostalgia, a sense that I was returning, even though I had never been there before? Some such thoughts came to me as I stepped on the riverboat.

The boat (*Dong Fang Huang Di,* the East King or King of the East) weighs 5,500 tons, can accommodate two hundred passengers, and has definite pretensions to luxury. Early in the evening of June 9, the Zhu family and I walked up the plank into a chandeliered, circular foyer. A disciplined staff of young male and female attendants, all good-looking and smartly uniformed, showed us to our cabins, even though A-Xing and Samuel, having been there before, already knew where they were. We had to go through our choreographed paces, and I must say I enjoyed the social dance— indeed, I enjoy almost all social dances in which I play a minor but dignified part. Each cabin, though small, was clean and comfortable, equipped with bed lights and table lights, telephone, toilet, wash basin, and shower. The East King was scheduled to remain moored until the middle of the night when, with all passengers asleep, it would set sail.

First Day on the Yangtze River

Pure Happiness

I woke up a little after six and looked out of the porthole
to see the Yangtze River flowing by and beyond it green
hills. For a moment, I was flooded with the wonder and
pure happiness of a child.

What happened? My guess is that a set of
circumstances came together that is unlikely ever to be
repeated. Having slept well—a rare occurrence at my
age—played a role, and that itself requires an explanation.
That first night on the riverboat, I tucked myself in
between the clean sheets of my small but comfortable
bed and twitched a knob by my bedside so that the lamp
gave rise to a warm orange glow. I smiled at the thought
that I had passed the midpoint of my journey and that
the rest would be easy since no speaking engagements
lay ahead; I realized, finally, that I could be irresponsible,
that I didn't have to worry about buying plane tickets,
checking in and out of hotels, or even hailing a cab.
Because A-Xing had clearly taken charge and was going
to mind all the harassing details of travel, I could be as
free and easy as Samuel and Alex. Samuel and Alex
didn't carry any money. I, an adult, went back to China
loaded with traveler's checks, Chinese yuan, American
dollars, and coins that distended my coat pocket, but

waking up on June 10, I finally realized I didn't need any money, that in the paradise of my second childhood I had no use for it.

Children, without a care in the world, sleep well, and I did too for the same reason. Lying in my bed in a riverboat, at rest and yet moving forward to a preset destination, was a new experience for me. I wallowed in the sensuality as a child might. But there was something else that morning that required a degree of maturity, namely, an awareness that I wasn't floating on just any river, that I was afloat on one of the great mythic rivers of the world, that only waking up on the Nile or the Euphrates could induce in me a comparable degree of wonder.

I was eager to get up for breakfast. That anticipatory delight in eating was another sensation I associated with childhood. Of course, I have known hunger in old age, but it was too often the churning of stomach acid rather than the urgency of a healthy appetite. The boat schedule announced breakfast at eight. I envisaged a feast of congee with "thousand-year" egg and Danish pastry with coffee, and I was not disappointed.

Something else I looked forward to was pure leisure that day and the next two days on the river. That didn't quite happen. I was on a tourist boat after all, and tourists, like small children, need to be constantly entertained, or so the management of the East King believed. I picked up the schedule of the day and saw listed taiji (tai chi) morning exercises, demonstrations of Chinese massage, acupuncture, landscape painting, snuff bottle painting, and a visit to the Shibao (Precious Stone) Pagoda. I yielded only to the last enticement.

Trying to Be a Tourist

I went on the pagoda tour reluctantly. A common bugbear of academics is to be thought superficial—mere tourists, gatherers of information that is superior only quantitatively to that which might be found in a brochure. My brochure said of the pagoda that it "was built in 1662 during the Qing dynasty. It is a wood structure with glazed tiles and red walls. With the whole body attaching on the sheer cliff, the 56-meter high and 12-caved pagoda gets increasingly narrower to the top. Without using a single iron nail, the pagoda is regarded as one of the eight wonderful constructions in the world and a brilliant pearl on the Yangtze River."

It was humbling to think that I would have nothing to add even if I joined the other tourists and climbed "one of the eight wonderful constructions in the world." Only a prior immersion in history could have prepared me to see more and more deeply. I had not done so. But what's wrong with being a naive tourist, open to the richness of a new place like a child? Nothing, of course. I am pointing out here the pride and vanity of a man with pretensions to learning against the innocence and humility of the ordinary tourist. The one, equipped with detailed knowledge, may come to believe that he actually bestows meaning and importance to place. It could seem to him that place in itself is essentially inert and mute until he comes along and gives it life. The tourist's attitude is the reverse. He goes to see Shibao Pagoda, or some other famous site, because he believes that it can enrich him. The same humility may make him eager to stand next to a celebrity and, if possible, to have a photo taken with him.

These thoughts darted through my mind as I put on my tourist's hat and got off the boat along with other people. On shore, we immediately ran the gauntlet of hawkers, who tried to sell us a choking variety of merchandise. The number of street vendors had swelled in the past ten years, in part (I was told) because peasant farmers had been displaced by the river's steadily rising level in response to the construction of the Three Gorges Dam. Peasants had become merchants. I wondered what they thought of the change. Life on the farm was undoubtedly harsh, a constant struggle against nature. Life as a small merchant and hawker couldn't be easy either. The struggle in this case was against rival vendors. So many voices were raised and so many hands reached out that clearly one person's gain was another's loss. Farmers cooperated with each other. They had to. Did the small merchants have a strategy of buying and selling, of keeping an eye on one another's goods, that required cooperation?

I climbed up the winding, slippery slope to the entrance of the pagoda and stopped. I was out of breath and didn't think it wise to go farther. A-Xing stopped, too. He helped me down. We once more had to run the gauntlet of hawkers. This time we appeased them by buying chocolate-coated ice-cream bars. I felt strange sucking on mine. Shouldn't I be eating a sesame bun? I wasn't the only one to have returned to the boat. A few other oldsters did the same. The hardier and younger tourists—including Ouya, Samuel, and Alex—continued their pilgrimage.

Second Day on the Yangtze River

Reflections on Labor and Work

A side tour up the Shennong River, a tributary of the Yangtze, first by ferry, then by sampan, was the scheduled adventure of the second day. Given my poor showing on the previous day, I decided not to go. What if I had to "go to the bathroom" while on the sampan?

All guests disembarked, including the Zhu family. For the next couple of hours, I had the riverboat and its efficient staff to myself. I was to know briefly what it meant to be royalty, alone in my palace, my every wish catered to by hovering servants. I luxuriated in the bright space, the sunlight pouring through the windows, the privacy and the silence, the stillness of a boat at anchor. I sat at a little table on the balcony that encircled the foyer and started to record the events of the last few days in my journal. Throughout this trip to China, all events seemed to me memorable, for all were touched by novelty, even such commonplaces as hailing a cab and eating in a restaurant. So I was puzzled by the difficulty I had recalling them. I suppose if all events stood out, then none did.

I knew that the tourists were about to return when male and female attendants converged on the foyer with hot towels and tea. They formed two lines so that the passengers walking up the gangway and then between the lines could receive the refreshments and imagine themselves intrepid explorers, tired but happy to be back. Alex, the quiet one, told me that the sampan was pulled up the swift-flowing Shennong River by men on the bank, and that they sang work songs as they pulled. I asked Alex what they wore. They had on shirts, shorts, and sandals. Well, I thought to myself, this trip up the Shennong on the sampan was meant to provide tourists with a feeling of what it was like in the old days, when men regularly pulled boats upstream against the swift current of the Yangtze. But there was a difference. In the old days, the men worked naked. I knew because in the two hours I had to myself, my stroll up and down the length of the boat took me through a corridor lined with framed pictures. Almost all were photographs of landscape scenery, except two, and these showed—most incongruously—men as naked draft animals.

Working naked, whether in the coal mines of Britain or on the narrow paths of the Three Gorges, seems to me the ultimate humiliation. True, the men were not commanded to dig or pull naked; they did so because it was, for them, relatively comfortable and convenient. A degree less humiliating than working naked is its opposite: working dressed up as fashion dolls. Such was how I saw the attendants of the East King riverboat. They were, as I mentioned earlier, all young and physically attractive. In cleanness and polish, they matched the scrubbed marble floors and waxed balustrades. Hygiene no doubt was demanded of them, and this meant frequent showers and many changes of underwear in the staff's dark, cramped quarters, and regular inspections of fingernails and breath. Speech, too, was scrubbed clean of dialect and made genteel. As for costumes, it was as if a spoiled child were in control, changing one set to another on her dolls as her mood dictated. In our three days on the boat, I was to see the standard uniform, ethnic dress, pirate's gear (complete with black eye patch), and the brocade robes of emperor, empress, and courtiers. The staff might wait at the table, make my bed, and clean my toilet, but they were also required to dress up and put on amateur performances for my and the other guests' amusement.

The Three Gorges

Passing through the Three Gorges and then through the five locks—the one a natural colossus, the others manmade giants competing against each other for supremacy—was the peak offering of the entire boat ride. Given all the advanced billing and my awareness of the impact of rising water, I had expected to be disappointed by the natural colossus. But no. The gorges and the mountains that towered over them cast their magic spell, and I was lifted, despite myself, to another realm of being. Certain natural features in the world—for example, the Grand Canyon and Yosemite in the United States—exude such majesty that no amount of prior knowledge can diminish them.

The impact of knowledge on perception is ambivalent. If knowledge is in the form of tourist brochures, postcards, and cheap photos, it can produce a feeling of familiarity, totally unjustifiable, that damages perception. But if it comes as history, legend, poetry, and landscape art of a high order, the effect can be enhancing. So it was for me, even though I had only a meager knowledge of the battles that had been fought in the Three Gorges at the time of the Three Kingdoms (220–80), the barest acquaintance with T'ang poetry and with paintings through the ages that paid homage to the ethereal beauty of mountains, waterfalls, and mist.

Something puzzles me here. I understand that the poet and the landscape artist can make nature more meaningful by infusing it with the poignant music of humanity. But would historical events, the noise of battle, be able to do the same? Wouldn't they have detracted from nature? The answer is, not inevitably; not, for instance, if the sounds of clashing swords and neighing horses were distant and faint, merging with the sounds of soughing wind and chattering monkeys. A natural colossus like the Three Gorges—unlike peach blossoms that come and go—lacks poignancy. Maybe the ghostly memory of ancient battles and passions gives it a touch of the bitter sweetness of the transient.

The Dam and the Locks: The Yu Legend

As the East King glided toward the first lock, another thought came to mind. With the construction of the mammoth dam, even the Three Gorges can be recognized as a creature of time. They endure as few things endure, but they are still vulnerable to human interference. Even if their basic structure remains intact, their appearance can be changed quickly in major ways. When the waters back up to their full height, the cliffs will no longer soar, and the once rapidly moving river will become a placid lake the length of California.

Power, whenever and wherever it is obtained, is used against nature and other people for glory, but also for practical ends such as transportation, flood control, and defense against human enemies. In regard to flood control, one might say that there is a precedent for the Three Gorges project in the legend of Yu, conqueror of flood and founder of the hereditary Hsia dynasty. In one version of the legend, floodwaters rose and threatened widespread destruction. The people were miserable. Their groans were heard by the Supreme Lord, who commanded Kun, father of Yu, to tame the flood. Kun attempted to dam the waters and failed. Yu was asked to continue with the task. Rather than trying to contain the waters, Yu dug channels to drain them off to the sea. His success meant that people could safely live on the land and cultivate it.

The Yu legend might not have much basis in fact. Its popularity suggests, however, that even in earliest times the Chinese admired mammoth engineering projects, provided they were carried out to benefit the people. That nature must be controlled and that in controlling it damage might be done did not seem to have distressed many Chinese then, nor does it do so now, much to the chagrin of Western environmentalists. Unlike Europeans and especially Americans, Chinese people have not embraced pristine nature as an ideal. The celebrated scenic spots of China—Huang Shan, Tai Shan, the surreal limestone towers of Guilin, and the Three Gorges themselves—all disclose religious architecture and are, moreover, soaked in legend, history, and art. Chinese landscape painting itself provides evidence of a lack of commitment to pristineness. Well known for showing nature in all its wildness and sublimity, yet it is never without some sign of human habitation and use: a hut here, a fishing boat there, or a traveler on a donkey moving up a mountain path.

Steps answer human needs. See steps anywhere and one can infer the human presence. Strange that I should be more aware of them in China than in any other country that I have lived in. My own increasing frailty is undoubtedly a factor. It seemed that everywhere I went there were steps to be negotiated—into restaurants and shops, at the Great Wall and at almost all the temples and pagodas, everywhere in the hilly city of Chongqing. And now, as our riverboat moved through the locks, I saw that they too were steps, giant concrete steps built on the Yangtze to allow boats our size to move up and down the river.

Steps help us walk on steep inclines. They are uniquely suited to our upright, bipedal posture and motion. Quadrupeds and other primates probably find them more a hindrance than a help. Steps are also a device to contain the pull of gravity. They extend the area of arable land and, at the same time, slow the downward creep of soil. China is an agricultural civilization. Not surprisingly I see steps—terraces—on hillsides everywhere. Floodwaters rush downslope, too, and dams are built to prevent them from doing so. And that is another reason for the Three Gorges project, namely, to control floods, by damming and digging, as the legendary heroes Kun and Yu did.

Stopover in Yichang

The riverboat attendants put on their final fancy costumes to perform the Dragon Dance as we passengers disembarked at Yichang. We were about to leave our world of affluence for one of poverty, for on the dock, we encountered porters, both male and female, gathered in large numbers, shouting and waving their arms to catch our attention. I was amazed to see that a porter, with just a short stick and strands of rope, could carry so many suitcases and bags. I was also pained that my fellow humans should fight for this backbreaking job. The five of us—the Zhus and I—sought to get into one cab, even though the law allowed only four passengers. When we tried to persuade the driver to discount Alex on the grounds that he was only a child, the driver retorted, "So a child is not a person? Tell that to the police!" On our way to our hotel, A-Xing stopped at a place that looked to me like just an ordinary store to buy airplane tickets for our next destination, Shanghai. Again it was A-Xing who faced the world. Ouya, the children, and I waited in the air-conditioned car. We waited ten, twenty, thirty minutes. Many forms had to be filled out, a further complication being that, though we traveled as a group, our passports were issued by three different countries.

Coffee at the Peach Blossom Hotel

I expected Yichang to be a rather sleepy river town, but, no, it too presented a skyline of tall buildings and cranes, and streets chock-full of cars and trucks, zigzagging in and out of traffic lanes as though they were in some sort of drag race. The day was warm and muggy. Our hotel, the Peach Blossom, offered a blessed oasis of coolness and calm. Bearing in mind that we had an early-morning flight to Shanghai ahead of us, we decided to rest that afternoon rather than attempt any sightseeing. At about three o'clock, however, A-Xing and I left our respective rooms and went down to the lobby. There we sat, drinking coffee and allowing our conversation to drift in and out of topics as the spirit took us. I mention this because one of the delights of this trip was that there were these laid-back interludes between new experiences and adventure.

At the Airport with Friends

Sitting alone in an airport waiting room has always been a sort of hell for me. Will the plane depart on time? Will there be postponements with no explanation? Will the flight be finally canceled, stranding me in a world of strangers, also uncertain of their fate? My anxiety can rise to a state of existential angst, making me wonder why I am making the trip in the first place, and even, neurotically, why I bother to continue with the journey of life itself. But when I travel with some-one, which happens rarely, I am a different person. I take delays, even cancellations, in stride. Of course, it helps to have someone with whom to talk over the delays, but there is more to it than practicality—the idea that two heads are better than one. It is rather the feeling that being with a friend at whatever place, even an airport, is an end, a sufficiency, in itself. With someone by my side, I am already in place, and so the question of whether I can get to another place seems less urgent.

There we were, the Zhus and I, at the Yichang airport. The time was 6:00 a.m., and we had another hour to wait before our plane's scheduled departure for Shanghai. We ate a leisurely breakfast. Rather than anything Chinese, I ordered coffee and a sweet roll, foods better suited to the airport's bland, international atmosphere. We then moved to the waiting room close to the boarding gate. I stretched out on my chair with Samuel on one side and Alex on the other. Ouya sat opposite us. A-Xing went to the bookstore to check what it had to offer. As the loudspeaker announced the boarding of our flight, he returned and gave me a CD of classical Chinese pipa music. He no doubt thought it would balance the Beethoven and Mozart CDs I had brought along.

Shanghai: Old Memories and New Experiences

August 1937–July 1938; December 1946

My family lived in Shanghai between August 1937 and
July 1938. I was six to seven years old. I can't remember
much from that period, yet certain images have stayed
with me for reasons I cannot fathom. In one, I am
sturdily refusing to hold my cousin's hand, as we children
were told to do, in a group photograph of eighteen
cousins who all happened to be in Shanghai in 1938.
Normally a well-behaved and obedient child, why did
I refuse? I was dissatisfied with something, and it could
be that I didn't like to be placed at the outer edge of the
group when my younger brother, who sat in a high chair,
occupied the center. Another image is of my brothers
and me standing outside our home to watch a dogfight
between Chinese and Japanese airplanes. They wove in
and out, spitting fire at each other, as though for our
entertainment. A third image is of a shopwindow display
of chocolates made in the shape of cute animals. When I
was given one, I was torn between wanting to eat it and
wanting to keep it as a toy. The fourth image is of the
child film stars Shirley Temple, then at the height of her
fame, and Bobby Breen, the eight-year-old boy who sang
"Buy My Flower" in the movie *Rainbow on the River*
(1936). A film that I guiltily adored had older boys

fighting one another with stones. Their athleticism and courage gave me an erotic charge that I couldn't then remotely understand.

To these early images in Shanghai I can add ones I garnered in 1946, in a two-week layover on our journey from Manila, the Philippines, to London, England. I was then a vain boy of fifteen, proud of my schooling overseas, my fashionable clothes, and especially my new bicycle, which I rode showily on the suburban streets of Shanghai. My parents, sister, and I stayed in the home of Father's close friend. He had a son my age to whom I made shy overtures, but to no avail. I was put in his bedroom. He made a point of never entering it until he could be sure that I was already in bed and sound asleep. Another memory from that time embarrasses me to this day. Father liked to boast of my English-speaking ability. One day, after dinner, he asked that I describe our journey from Manila to Shanghai for his friends and to do so in English. I was happy to oblige, for I ignorantly thought I knew the language well. But I didn't. I stammered and stuttered, and committed all sorts of solecisms that even I could recognize at the time. I was overcome with shame.

The View from My Window

And so here I was, back in Shanghai, an old man. How satisfying it would be if I could honestly add but "no longer proud and vain"! The Zhu family and I checked into the Seagull Hotel. From my tenth-floor room, I had the most splendid view of downtown. Straight ahead was the Huangpu River; to the right, the famous Bund of solid colonial buildings—banks, financial institutions, customhouses, and hotels; to the left, a fantasy of postmodern architecture that included the Oriental Pearl TV Tower (a slender spire that threaded two spheres) and the eighty-eight-story vertical pagoda of the Jin Mao Tower. At night, both banks of the Huangpu were brilliantly illuminated, as was the river itself, shimmering with lights from the shores and from showboats that were webbed with colored bulbs.

From my hotel room, I could also see, early in the morning and late in the afternoon, crowds of men and women performing Mao exercises in the public square. Such disciplined group activity seemed to me eminently Chinese: true, it was promoted by communism, but in this case, communism merely reinforced the Chinese proclivity to do things together. That was not new, but the sight of men and women jogging alone was. Was this another sign of rising individualism?

Beyond the foreground of milling people and motorized traffic, beyond the parade of boats on the Huangpu River, beyond the midground of staid and Disneyesque buildings, was what? I strained to see but couldn't at first decide whether I saw a distant range of mountains, a bank of dark clouds, or a row of tall buildings. It was tall buildings! Some three thousand of them were built in a mere fifteen-year period. Shanghai now has more high-rises (defined as a building of more than eighteen stories) than New York City. Construction so rapid and on such a scale generates a feeling of unreality, for it is hard to imagine all these high-rises more or less occupied and the people in them going about their routine business. Looking out from my hotel window, I had to convince myself that the distant skyline was not just a movie mogul's facade or the clever effect of digital imaging.

Dinner on a Boat Restaurant

That evening, the Zhus and I had dinner in a restaurant designed to look like a row of docked boats. Our boat was illuminated by lanterns that glowed hospitably in the darkness of night. We sat in comfortable chairs placed along the two sides of a long table. "Hey, Alex, are you going to order fried rice and Sprite?" I asked teasingly. We were in a relaxed and slightly wistful mood near the end of our journey together. I was to return to the States in two days; A-Xing was to go to Beijing to prepare for a conference on fuzzy logic; Ouya and the children were to go to Ouya's hometown and spend some time there.

We chatted merrily. But what did we talk about? No doubt food was a central topic. What else? If I now offer an example of our conversation, it is because of my belief that the level and tone of what we said to each other gave our little community its unique flavor and spirit. I knew Samuel was good at math—I must have brought up the topic of his sweater again, the one that had the logo "Future Engineer of Wisconsin" printed on it—so I asked him whether he was also good at deductive logic, at deduction. That word didn't ring a bell. So I asked, "Have you heard of Sherlock Holmes?" He had. I wasn't surprised, for I knew that Samuel read widely. Still, I was pleased that I could share this childhood hero of mine with a child of the twenty-first century. We had almost finished eating. Hunger assuaged, it remained for us to pick up our chopsticks now and then to secure a mushroom or a prawn and put it absentmindedly into our mouths. This was the best time for a raconteur to show his stuff. So I said, "Let me tell you a Holmes story to illustrate the power of deduction." (I told it in English, in part to accommodate Samuel, in part because I was mentally preparing myself for reentry into the English-speaking world.)

"Sherlock Holmes and Dr. Watson went camping on a desert plateau. In the middle of the night, Watson was rudely roused from his sleep by his companion, who asked, 'Watson, look up and tell me what you deduce.' Watson looked up at the night sky. He saw its brilliant stars and offered Holmes a brief account of the Big Dipper. Holmes shook his head and said that he asked for deduction, not mere description. So Watson tried again: 'I see wisps of cloud blanketing the stars in the eastern sky. I deduce that it will rain tomorrow.' Holmes shook his head and said, 'Plausible, but irrelevant.' The patient Dr. Watson tried a few more times without success. Finally, he said, 'Well, what is the correct deduction?' Holmes replied, 'Someone stole our tent!'"

Samuel burst out laughing, as did his father. His mother smiled. Only his brother Alex looked a bit uncomprehending. I judged my story a success, as indeed were the dinner and the entire magical evening. It just might be, I thought at the time, that no other table was more cosmopolitan than ours in a city that took pride in its cosmopolitanism. Here we were, citizens of three different countries, eating a Chinese dinner, speaking in

a mixture of Chinese and English, and admiring the
analytical powers of a detective conjured into being by
Sir Arthur Conan Doyle.

A Stroll along the Bund

A-Xing suggested that we take a short stroll along the
riverbank before turning in. Even at nine o'clock in the
evening of a regular workday, the sidewalk was jammed
with people enjoying the cool night breeze, the floodlit
buildings on both sides of the Huangpu River, and, above
all, one another. Friends and lovers stood still for digital
cameras, disregarding the seething crowd around them.
Winding through the crowd were rivulets of tourists,
each led by a pennant-carrying guide who shouted to be
heard, dispensing scenic information and warnings to his
charges to stay together and move along. The accents of
the guides indicated that the tourists were out-of-town
Chinese, some from less developed provinces. There they
were, gawking at Shanghai's sophisticated night scene,
no doubt surprised that this, too, was China.

On our way back to the hotel, we saw a legless beggar.
He was the only beggar I saw on the entire trip, which is
amazing when I think how the streets of Shanghai
thronged with them in 1946; and amazing, too, when I
think that every day I walked State Street in Madison,
on my way to school or home, I would be stopped by a
panhandler every two to three blocks. They might be
in urgent need, perhaps for a drug fix, but they looked
reasonably intact and capable. The Shanghai beggar, by
contrast, was a truncated being—severely handicapped
in the struggle for life. A-Xing put some money in his tin

cup. Shamefully, rather than follow his good example,
I was preoccupied with the thought that the difference
between the Madison panhandler and the Shanghai
beggar's physical state was the difference between a rich ·
country and a poor one.

Last Day: Food Poisoning and Conversation

Luxury in Sickness

I woke up in the early hours of the morning, feeling ill. I rushed to the bathroom and vomited on the floor and on my T-shirt before I could quite reach the toilet bowl. Again and again, I had to bend over it. At seven, as we had agreed on the previous night, A-Xing called me from his room to ask whether I was ready for breakfast. When I told him about my throwing up, he said that Samuel was similarly afflicted. Both of us had had several servings of "dragon-well prawns," which were undercooked but delicious—delicious because undercooked. They must have been the culprit.

I slept through the entire morning with periodic visits from A-Xing and Ouya, who took such good care of me that, when the nausea abated, I was able to luxuriate once more in the sweet "letting go" of childhood. A-Xing consulted the hotel's resident medic, who wasn't much help, except to say that I might check in at a hospital that catered to foreign tourists. A-Xing offered to take me there. When I rejected the idea, he urged that I rest, that I sleep if possible, and not be concerned with tomorrow's flight back to the States, for he could arrange to postpone my departure by a day or two should that be necessary, and that, in any case, the final decision could be left till early the next morning.

Conversation in the Lobby: Cultural Differences

One scheduled event remained to be taken care of on my last full day in China, and that was to meet with Professor Zhou Shangyi of Beijing Normal University. Zhou Shangyi and I had gotten to know each other fairly well in her year as a visiting scholar at UW-Madison. It was she who made detailed arrangements for my visit to her university. She was flying in from Beijing so that, before I left for the United States, we could make one more attempt at resolving certain difficulties in the translation of my book *Escapism* into Chinese. She also wanted to introduce me to her friend, Gang Zeng, dean of the School of Resources and Environmental Science at Shanghai's East China University.

At ten in the morning, when Shangyi's plane took off for Shanghai, I was still far too sick to contemplate lunch. A-Xing offered to host our visitors at a restaurant, and from there to call me at the hotel to see whether I was well enough to have company. At one in the afternoon, I had recovered sufficiently to tell A-Xing that I would wait for them in the lobby.

I sat near the revolving door and kept an eye on the people coming in. Suddenly there was A-Xing. He rushed in ahead of his guests, waved at me, and said that he too was beginning to feel the effects of food poisoning, that he had to go to the washroom, and that he would join us later. I greeted Shangyi, who introduced me to Gang Zeng. He asked whether I was feeling better, which I suppose any polite person would, but he followed it with the offer to take me to the hospital. That extra concern caught me by surprise, for he must have surely known that, had I agreed, he would have had to fight city traffic and waste an entire afternoon with me.

A-Xing joined us. We talked about the challenges of translation. Shangyi was keen that the Chinese rendition of *Escapism* should have literary merit in its own right. Of course, there were the usual difficulties of making words and syntax in one language convey the meaning of words and syntax in another, obstacles that couldn't be overcome completely. There was, however, also a more superficial difficulty—a cultural difference between China and the West as to what is and is not acceptably said in scholarly writing. In *Escapism,* I include descriptions of the sexual experience in plain English. No Western scholar or reader will find them objectionable, but the Chinese still maintain a fairly sharp distinction between "high culture" and "low culture": the one carefully maintains an elevated tone and avoids the use of certain words and expressions, while the other freely traffics in the lurid and the obscene.

I asked Gang Zeng, the dean, what his academic specialty was. It was economic geography. After some prodding on our part, he gave a fascinating sketch of Shanghai, its economic base and global aspirations. I saw once more how educational and entertaining telling others about one's area of expertise ("shop talk") could be, and how its prohibition in social settings can make conversation so insipid. When it was my turn to say something, I briefly encapsulated the principal thesis of my book *Passing Strange and Wonderful.* It emphasized the importance of the aesthetic in almost all spheres of life, including the economic and the political.

I became rather excited and probably spoke too long.

Two hours had passed since we had all met in the hotel lobby. Nausea threatened to recur and I begged to be excused. Back in my hotel room, I wondered whether I was going to vomit again. But no matter. Physical stress counted for much less when it wasn't also burdened by mental or psychological anguish. I felt surprisingly at ease and at peace, reassured (perhaps) by my recognizing another cultural difference between China and the West. In China, people were willing to give a scholar credit for his labors over a lifetime and did not insist on the adage "You are only as good as your last paper." I realized that I didn't have to be always on my toes, straining for one more (perhaps impossible) flight.

To the Airport
and Home

If A-Xing had not called me at 6:30, as he promised, I
would have slept soundly on, a sign that I was free of the
worst effects of food poisoning. The cabdriver told us
that, before the construction of the freeway, it would
have taken us two to three hours to get to Pudong/
Shanghai International Airport. The time now was more
likely to be three-quarters of an hour. "But, of course,
if you really like speed," he added, "the speed train—
the fastest in the world—would have taken only seven
minutes, though you might need an hour to get to the
train station." He said all this with a mixture of skepticism
and pride. He could have been a talkative cabdriver in
New York. As I listened to him, I was amused by the
thought that values traditional to America—size, height,
and speed—have been taken over by the Chinese, or
rather, by the dynamic citizens of Shanghai.

The Pudong/Shanghai airport is huge and forbidding.
My head swarmed in confusion as I looked at the giant
schedule board packed with the times of arrival and
departure in Chinese and English. Fortunately, I only
had to pretend to look, for I counted on A-Xing to do
the real looking and tell me where to go and what to do.
A succession of Kafkaesque procedures led me finally to
the security checkpoint, beyond which A-Xing could not

go and I would be, for the first time in more than two weeks, completely on my own.

Chicken for Dinner—Western Style

The long flight back to the States was blessedly uneventful. I still worried about the state of my stomach. For dinner, we had the choice of Western or Japanese. I chose Western. On my little table, the whitest tablecloth, cool glistening silverware, a sparkling slender-stemmed glass, and a milk-white china cup that looked as though it had just emerged from a bubble bath were all a beneficence to one who still felt a little queasy. But despite the big comfortable chair, despite the leisurely meals that took up a chunk of time, the eleven-hour flight from Tokyo to Minneapolis was a trial. I tried to make it go faster by switching on the home movie—not the little computer screen in front of my chair, but rather the one in my head, newly refurbished with reels from China.

What? No "Welcome Home"?

Soon after setting foot in my adopted country, I met with my first big disappointment. The immigration officer in Minneapolis failed to say, "Welcome home!" Ever since I became a citizen in 1973, I looked forward to stopping at the booth where, upon presenting my passport, the officer would acknowledge my right to be where I was with these two gracious words. The first time it happened was in 1975. I was returning after a year as a Fulbright scholar in Australia. I walked toward the booth with the trepidation of someone used to being treated with suspicion, an alien. But this time, would he recognize me as a citizen? He did. On subsequent returns to the United States, the officer's "welcome home" still worked its magic even though it no longer had the element of surprise. I suppose the reason for the greeting's staying power is that although I enjoyed my visits abroad—to England, Denmark, and Australia, for instance—I was never wholly at ease. The tension in my body started to ease only when the airplane finally touched down at one of the usual points of entry—New York, Chicago, or Minneapolis. I had come home, and I liked to have that feeling officially confirmed. But is "returning" the right word this time? And if I say that I have come back to the United States, is "back" the right word? Shouldn't I reserve them for China?

The final leg of the trip, from Minneapolis to Madison, was just a short ride, barely long enough to be served diet Coke and pretzels. Our plane landed at Dane County Regional Airport at 4 p.m., exactly on time. I walked along the airport's long corridor and then down a flight of steps to the waiting area. And there was Qiguang, one of A-Xing's graduate students, waiting to welcome me back to Madison, my home for the past twenty-two years.

Reflections

Welcomed home by a Chinese student! Don't I have the right to expect an American student? I know the wrongheadedness of the question, for I never sought to be met: I had made no prior arrangement because, after all, Madison is home; I am familiar with it, I don't need any help. A-Xing had seen how sick I was, however, which was why he had asked Qiguang to meet me. I was happy to see him, happier than I thought likely. That could only mean that, in the depth of my being, after a journey abroad of this length I yearned to see a welcoming face, to feel that I was back where I belonged.

But where do I belong? Am I a Chinese, an American-Chinese, a Chinese-American, or an American? There are two sides to this question, one impersonal and the other personal. As an impersonal question of taxonomy, where one belongs is worth considering if only because many people are confronted by it in our increasingly pluralistic society. In responding to the personal aspect, I remain at loose ends for an answer. In this regard, the return trip to China has not helped. It has done almost the opposite by stirring hopes and longings that should perhaps be left in the bottom drawer of my psyche.

I will conclude this journey into self and culture by taking up both sides of the question. As a student of humanistic geography, I can hardly avoid wondering how

culture and self mutually sustain each other, but my
concern over the issue had remained rather abstract and
academic until, several years ago, I read about a football
match between Americans and Iranians in Los Angeles.
It was an early attempt at breaking the ice between the
two countries, somewhat along the lines of the "ping-pong
diplomacy" that was said to have brought America and
China closer together. The huge crowd in the stadium
that day was largely Iranian or of Iranian descent, and
included many young men who were born and raised in
Washington, Oregon, and California, and had lived in
no other country than the United States. Whom did they
root for? They rooted overwhelmingly for the Iranian
team. I wasn't really surprised, but it led me to ask, if the
game had been between Americans and visiting Chinese,
where would my loyalty lie? How would my emotions, an
unfalsifiable index of deep-seated loyalty, respond?

Belonging at Different Levels

People need to feel that they belong. At one level, how
surprisingly easy it is to evoke the feeling. In a football
game, wearing a uniform and adopting a mascot suffice.
Minnesota players wear maroon and are Gophers;
Wisconsin players wear red and pretend to be quite
another kind of animal, Badgers. Thus differentiated
they are ready to play. In actuality, members of
both teams are very much alike; they are all young
Midwesterners of European and African-American
descent. The identity they acquire in the game and the
loyalty they feel for their team are cultural markers that
they can put on or take off.

The identity I am wrestling with lies at a deeper level. Their cultural markers—clothes, house type, food, music, dance—are things of everyday life, not things that people can put on or take off as in a play or a game. When we speak of a global culture, we are saying that people worldwide want the same sort of commodities in the conduct of everyday life, and that, this being so, their psychology and way of thinking must also be more and more alike. What people make every effort to get, because they seem necessities rather than passing fancies, will sooner or later define who they are.

In China, my initial cultural shock is diminished by the Western style of dress adopted by nearly everyone, in particular, the young. Students at Beijing Normal University look and act much like students at the University of Wisconsin: they wear the same sweatshirts, jeans, Nike shoes, and carry the same backpacks. But the similarity goes much further than attire. Chinese students patronize Western-style fast food; they like Western popular music, and they even rather enjoy the novelty of speaking English. And yet, once I get to know them, their "Chineseness" comes through. It comes through as certain values, for example, respect for the old and an appreciation of quietude (manifest in people silently going through the movements of tai-chi in a public park), that have survived both the fanaticism of the Cultural Revolution and the present headlong drive for wealth. These values, like material goods, are public in the sense that one can see them in gestures and behavior. Beyond them—or perhaps I should say, underlying them—is patriotism and civilizational pride, a feeling or mode of being that rarely rises to surface consciousness and is the opposite of flag-waving. Its ultimate source is history, geography, and language.

The Triune Roots of Identity

But what do I mean by history, geography, and language? These areas of knowledge have both a public and a private aspect, and it is important that we know this to be the case. They are public in the sense that they are school subjects that can be formally taught. When a people is unsure of its identity, one solution is to teach it its history and geography, and make sure that its language continues to be used. A collective sense of self may well become firmer this way, just as it does when ethnic dress, food, and dance are adopted or reintroduced. However, besides such formally imparted knowledge, history, geography, and language can also mean something deeper, taught but not in any standard or prescribed way. History is, then, stories and hearsay that one learns in passing in childhood and through eavesdropping on the conversation of adults; and it is routine participation in the historically grounded practices and rites of the tribe, not the mere putting on of a show, or the self-conscious mining and miming of the past to affirm one's identity. Geography is an intimate bond with place, knowing it at the most basic level through one's senses and movements, knowing it practically in the course of carrying out the daily necessities of life, and knowing it emotionally through the use of charged words and deferential gestures. Language can establish or sever a relationship, and in this capacity it complements facial expression and other bodily stances. But it is also the conceptualization and imaging of a world, an activity that is unique to the human species. Understood in both their public and private aspects, the triune of history, geography, and language undergirds a people's strongest sense of self. It also undergirds an individual member's sense of self insofar as that individual is integrated into the group.

Where Do I Stand?

Where do I stand in this triune? As a child, I was enthralled by the conversations that Father held with his friends in our Chongqing home. They ranged over every conceivable topic, from astronomy to Yuan dynasty drama; from world history—events of the First World War, for example—to whether a certain general in the Chan Kuo period had side whiskers; in geography from the wilds of Tibet to a cozy teahouse downtown; in literature from Ibsen to Lu Xun. I couldn't make much of what I heard, being only a child, but the lasting effect of staying up late and listening to adults was to imprint in me the idea that the world was vast and enormously exciting, that China was not only a part of that world but played a key role in it, that I was a Chinese and being Chinese I would have a place under the sun. I became very patriotic. I was patriotic for another reason. Father and his friends also talked about China's humiliation under Western powers and under Japan in modern times. Even as they chatted, laughed, and smoked cigarettes in the courtyard to escape the summer heat, Japanese airplanes might be coming over the horizon, forcing everyone to head for the caves.

At an early age, I did not so much formally learn as

imbibe an awareness that to be Chinese is to be at the
center of things and of the world. China's habit of
seeing itself as *Tien hsia* (that is, "all under heaven"),
"Within the Four Seas," and Middle Kingdom, somehow
got through to me. True, a mere century ago, almost
all people, no matter how small and isolated, saw
themselves as the only true humans before their delusion
was shattered by encounters with a more powerful
neighbor. Thenceforth, they struggled to maintain their
identity as one people among several, and even that
could be difficult since their original sense of self rested
on the belief that they were alone and, if not alone, then
certainly the brightest and the best. China's sense of its
own importance—its centrality—was hard to maintain
after a succession of mortifying defeats by small European
armies. Hard, but not impossible, for China's reputation
as a civilization, comparable in stature to that of the
West and of India, remained largely intact, thanks to the
advocacy of Western scholars. In European and American
universities, knowing something about Chinese history
and culture was considered a necessary provision of the
cultivated mind. Young men and women study Chinese
civilization in departments of history and literature, and
not in departments of ethnography and anthropology, as
they would if they were interested in the customs and
mores of nonliterate peoples.

In the United States, I am seen as a hyphenated
American. A hyphenated American is one who seeks to
have the hyphen dropped so as to become simply an
American. European immigrants have all more or less
successfully taken that step. The descendants of Africans,

transported across the Atlantic as slaves, have not and many could not. In time, they decided to embrace the hyphen, to be proudly African-American, and to use that pride to generate political cohesion and power. Other minority groups, struggling for acceptance with meager success, have adopted similar tactics. Not welcomed at the country club? Hard to climb the ladder of success in business corporations and universities? Minority-group leaders have argued both that these "glass ceilings" should be removed and, somewhat contradictorily, that their people should turn away from the values these institutions stand for and find, instead, strength in their own culture and traditions. Treated as an ethnic? Well, then, *be* an ethnic, be a people happily wedded to place, even if it is a sort of ghetto.

When I say I am a Chinese-American, an ethnic, I do so with tongue in cheek, for given my upbringing, I do not in the least see myself as a minority person at the margin of things. Do I see myself, then, as an unhyphenated American? Yes, but only because to my mind America itself is profoundly nonethnic, not a nation but many nations, not a people but many peoples able to come together, not only because they tolerate or even appreciate each other's customs, but, even more— and this is far more important to me personally—because they share certain universal ideals.

In 1951, when I was about to come to America for graduate school, my English friends seemed puzzled that I didn't prefer to live with them and be one of them, since I had received most of my education in Australia and England. "You will be more at home here," they said,

failing to realize that England at that time was too
culturally homogeneous, too intimately tied to that
"precious stone set in the silver sea" for me to feel an
integral part of it. America was and is different.
America is as much "space" as it is "place," as much
a possibility to be invented as a past to be savored.
American freedom is the freedom of the road, but at a
deeper level it is the freedom to be oneself, a unique
self in the midst of other unique selves, yet all, separately
and all together, able to recognize and rejoice in their
common humanity.

Sometime in the early 1950s, I realized for the first
time that I might feel at home in the United States. I
was still a graduate student and traveled widely over
California by Greyhound bus. After many hours on the
road and through the night, the bus stopped at a rest
station. We were to have breakfast there before moving
on to our next stop. I climbed out along with my fellow
passengers, all feeling rather groggy and stiff, and all
feeling a certain camaraderie for one another through
sitting close and even slumping over a neighbor's
shoulder in the course of the long ride. At the counter,
I ate a waffle with maple sauce and drank two cups of
black coffee. When I got out of the rest station, the
other passengers were also out there, standing near the
bus, chatting and smoking, waiting for the driver to
order us on board. Was it the waffle and the black coffee?
In any case, as I looked at the horizon of dark palm trees
silhouetted against a brightening sky, I felt a rush of
happiness—the happiness of one who was at once at ease
and excited. For the first time, I seriously thought that
I could be at home in America.

History

History—American history—was not something I picked
up through the accidents of overhearing grown-up
conversation. I consciously acquired as much of it as I
could through voracious reading. I took this step because
I wanted to know something about my past. My past?
Curiously, the answer is yes. Having decided to make
America my permanent home, its past became ipso facto
my past, my present, and my future. What I read in
books—the transatlantic crossing of the Pilgrims, the
slave trade, the Civil War, the settlement of the West,
and such like—did not seem to me just other people's
stories. I took them as my own because they impacted
my sense of self, altering and generally enlarging it. I
have had, after all, a similar experience in childhood. In
my Chinese elementary school, we learned about Isaac
Newton and his apple, Benjamin Franklin and his kite,
together with the doings of Chinese heroes. They were
offered to us without distinction of nationality, as
examples of what humans could achieve and therefore
what I, a Chinese child, could achieve.

I was naturalized an American citizen on December
19, 1973. In 1975, I went to Australia as a Fulbright
lecturer. Before departure, I received a form letter from
the State Department telling me that I, a citizen,
represented the United States and should behave with
seemliness and honor. I couldn't help smiling at the
notion that overnight my status had changed from being
an alien to being an ambassador for America. A reception
in my honor was held at the home of a professor at
Australian National University. We stood around, drinks

in hand, and chatted amicably until the topic veered to the war in Vietnam. Suddenly, the tone grew threatening. Australian academics were vehemently antiwar. One of them pointed his finger at me and said, "The trouble with you Americans . . ." I couldn't believe what I was hearing. Obviously I was an American to him. As obviously, no matter how much at home I felt in the United States by 1975, I did not feel a citizen with a responsibility for the massacre of Indians at Wounded Knee and the tragedy in Vietnam.

So, not a true American. Am I then a Chinese at heart? How deep is my attachment to China? I picked up odd facts about China's past at the soirees that Father held with his friends. I have emphasized the importance of this informal way of learning to one's sense of rootedness. Ironically, the stories I overheard in childhood were not restricted to China, but ranged widely. It was as if, at bedtime, an American parent told his child not only about George Washington and the cherry tree but also about the patriotism of the Song general Yueh Fei.

As an adult, I read books on Chinese history as I did books on American history. If I consider American history my own, wouldn't I consider Chinese history even more my own? That would seem natural. Yet I wonder, for printed text is a cool medium. A better test is this. In the United States, when I take a foreign visitor to the battlefield at Gettysburg, will I, when I explain it, use the pronoun "we" or "they"? If, in another scenario, I am in China and I take an American friend to the Great Wall, will I find myself saying "we Chinese" to him? Will I even find myself boasting or apologizing for what "my" Chinese ancestors have done in this part of the world?

Geography

Geography is another source of our rootedness. There is a certain mystique to being born in a place and growing up in it, feeling one's senses come to life and one's awareness expand in contact with the sights, sounds, and odors of a particular locality. In premodern times, one was literally fed by a patch of earth that was small enough to walk across in an hour, since nearly all one's foods came from it. Moreover, one's ancestors, one's community and its ascendants, were all fed and sustained by the same patch. "Where are you from?" is a common enough question even in mobile America, and when the stranger's reply is "Alabama" or "Wisconsin," one immediately feels that some important fact about him is offered. In China, the more specific question, "What is your hometown?" is routinely asked. I have learned to answer, "Yinshan, Anhui Province," because that's what my father told me. But Yinshan, for me, is just a place-name. I have never been there and know nothing about it.

I never got to know any part of China really well because my family and I were constantly on the move, escaping from the Japanese. In America, too, I never lived in any city for longer than five years until I moved to Minneapolis in 1969. And so if my personal experience of American places is superficial, so is my personal experience of Chinese places. Having said that, however, I need to make a qualification. I was a child in China. A child's openness to his milieu's sensorial qualities greatly exceeds that of an adult, dulled by routine and the chores of practicality. Chinese cities have therefore left their mark on me in tactile, olfactory, and visceral ways that American cities have not quite been able to do, even though I have lived in them much longer.

Now that I have visited China and have come back to the United States, what do I make of Chinese cities and scenic wonders? Has my book knowledge of them been altered or enhanced by the visit? To answer these questions, at least partially, let me turn to four places: Tianjin, where I was born; Chongqing, my last residence; and Beijing and the Three Gorges, places that I had not known previously in a direct and personal way.

Tianjin is just a fifty-minute drive from Beijing, but on this trip I couldn't fit it into my schedule. It wasn't just a matter of time, however. I think I also dreaded the thought of finding the place of my birth and young childhood totally strange, that in Tianjin I wouldn't encounter a single landmark that could resurrect a memory, that the sprawling, booming port city would wipe out what memories I do have. Two of these stand out. In one, I was taking my afternoon nap. Like most young children, waking up put me in a bad mood and I was inclined to throw tantrums. To prevent that from happening, my wet nurse did something very ingenious. She put water in an ashtray and then put the ashtray out on the window ledge, where the water, in Tianjin's cold winter, would freeze. As I was about to wake up, she retrieved the ashtray and knocked its edge against the top of the small table by my bed. Out came a brilliant, sparkling ice sculpture! I was stunned into silent wonder.

The second memory has a specific geographical location. It was Pa Li Tai, Nankai University. Father used to skate on a little pond in Pa Li Tai. He wore a thick sweater knit in such a way that a flap came up to protect the back of his head. I was a nervous four-year-old who didn't have much confidence in his father's skating ability, but was somewhat reassured by the protective flap. One day, as I watched, he fell backward, knocking his head against the ice. I cried inconsolably, believing that he had died. No matter how sophisticated my knowledge of Tianjin has become, these two images from childhood will always remain to give it a deeply personal flavor.

Chongqing, my last place of residence in China, has given me the most vivid images. Unlike other Chinese cities that I have lived in, geography figured prominently. I remember the dense fog that imparted a slightly menacing air to landscape; sizzling summer days when the greatest luxury was to bite into a cool slice of watermelon; drained rice fields carved on the hillside that became a gigantic stepped garden for children to clamber over; abandoned grave mounds in the midst of which we enacted war games; racing down Ko Lo Mountain with Father holding my hand and almost lifting me off the ground; dark and dank caves that we hid in during air raids; playing soccer in an alleyway at dusk before the clarion call for dinner.

Besides these place-specific memories, I recall two of a more general sort that, I believe, are widely shared among children. One is recovering from sickness, wallowing in guileless sweet rest, playing with toys that threatened to disappear in the billows of one's quilt, under Mother's watchful and loving eyes. The other is the sheer joy of being alive, inhaling the scented air of early morning by the lungful, racing across an open field as though one's feet were treading on air.

Naively, I thought I would be sucked back into the past when I revisited Jin Nan Tsun and Nankai School. That didn't happen. I should have known that to be alive is to be disloyal to the past. As I visited my old home and walked the school grounds, ghostly presences were pushed into the background by the sights and sounds right before me, and by nagging concerns of the moment. It was only after I had left Jin Nan Tsun and returned to my downtown hotel, and only now, two months later in the quiet of my study in Madison, that a feeling of tenderness emerged for my boyhood self and the environment that nurtured it.

Looking back at the places I have known, directly and indirectly, I am shocked by the arbitrariness of the ways I noticed, wondered about, or remembered things. I used to dismiss photo albums for their pathetically meager selection of images, and dismiss movies, too, for the huge gaps in time, for the way action jumps from one place to another. In such dismissals, I failed to realize that life as lived is also shockingly discontinuous, full of black holes into which a myriad of experiences and thoughts disappear, never to reemerge, except perhaps under accidental circumstances. How is it that, having lived in Chongqing for three years and having used the word "Chongqing" many times in speech and writing in later life, I have never stopped to wonder, "What do the two characters *Chong* and *qing* say?" They say "double celebrations." But why? Only now, after my return from China, have I learned the story behind the name, and that accidentally by flipping through the pages of a guidebook! The name was intended to celebrate Zhaodun's pacification of the area and his ascent to the throne as a Song emperor in 1189.

Did my appreciation of Chongqing seriously suffer from this gap in knowledge? I shouldn't think so. It is, after all, a rather nugatory fact. But that is the point. When I look at events in the world, I believe I can distinguish the important from the trivial, but when I look at events in my own life, I am not so sure. Often the small happenings—a hug at the right moment and the sudden awareness of the deep meaning of a word—stay in my memory and color my outlook on life, whereas the big events—those that are honored in the family photo album or appear on the front page of the newspaper—are forgotten like failed blockbuster movies.

Now, to Beijing. I was born in the neighboring port city of Tianjin. Sometime in the first five years of my life, I must have been taken to Beijing, but I have no memory of such visits. The five days I spent in the capital in 2005 were therefore my first real encounters with the city and its environs. How would they differ from those of any tourist? The principal difference, I believe, was my greater interest in the human and social as distinct from the purely architectural. On my second day in Beijing, I was given a quiz as we walked out of the hotel: "What do you think those tiled lines in the middle of the sidewalk are for?" I looked at the inconspicuous lines, puzzled. After one or two halfhearted attempts at an answer, I gave up. The correct answer was that they were put there to guide blind people as they felt their way with their walking stick. That's considerate of the government and the planners, I thought, in striking contrast to the heedlessness of the motorized traffic, which showed not the slightest concern for pedestrians, and in striking contrast again to the thoughtful individuals who guided me across the streets.

I looked for the red flags of the People's Republic. I expected them to be on display everywhere. They were plentiful in Tiananmen Square and they could be seen on government offices, but, in contrast to American practice, they were absent from commercial and domestic buildings. Does this mean that social cohesion in China is stronger and that there is therefore less need to make a show of the material symbols of unity? Another evidence of social cohesion—of the feeling of connectedness between people—lay in the parks that were put in the midst of high-rise tenements for mid- to low-income people. In New York or Chicago, urban parks tended to become scenes of drug dealing and crime. In Beijing, as I have already noted, people snoozed on benches or played chess, children shouted and shrieked on swings and slides, oldsters did their exercises on simple equipment painted in primary colors. This degree of harmony would be understandable if the inhabitants all came from the same villages and towns, but that did not seem to me likely.

While human relationships in Beijing aroused the greater curiosity in me, I was struck by one overall architectural change because of what it meant for the way people saw the world and conducted their lives. Old Beijing was a flat city that symbolically rose as a pyramid culminating in the T'ai-ho Tien at the center of the Old Palace. New Beijing—the metropolis I saw— was, in a curious way, the reverse. Many skyscrapers have been built, and, in this sense, Beijing is becoming a vertical city, but symbolically it is flat; that is to say, it is wholly secular.

On June 11, our riverboat, the East King, moved into the Three Gorges. I was there as a tourist, along with a hundred or so other tourists. How did I respond? Despite all the hype that might have induced a sense of letdown and despite the rising water level that diminished the height of the cliffs, I was impressed. With the passing of minutes and hours, however, my feelings toward this natural marvel became more mixed and contradictory. What they were seemed to depend on the direction I was looking and at the drift of my thought. Looking up, I saw towering peaks wrapped in mist, the inspiration of so much Chinese landscape art, but looking down, I saw a flotilla of filth bobbing on the muddy flow of a river that has been transformed into a vast sewer, China's cloaca maxima.

Nature at the Three Gorges, like nature at Arizona's Grand Canyon, is spatially immense and an abyss of time to those who can read the stratigraphic record. It dwarfs the human being into utter insignificance. Unlike the Grand Canyon, however, the Three Gorges are almost everywhere altered by the human hand and spirit. Terraced fields carved on even the steepest slopes and vast constructions like Gezhou Dam are the most obvious. But also altering nature—certainly the perception of nature—are the historical events that have taken place there, going back to 1000 BC, the battles fought and the temples built, and the stories and legends that have been told about every prominent peak and deep valley.

Can mere words damage nature's integrity? On the deck of the East King, I heard the guide urging us to direct our attention to that square indentation on the Red Rock Mountain, for "it was used by Lu Ban, the great inventor, carpenter, and patron saint of all wood workers and masons"; and to admire that stream coming

down to join the Yangtze, for it wasn't just any stream—
it was the Fragrant Stream, and the reason for its name
is . . . After the tenth or thirteenth story, I turned away.
I couldn't bear witnessing this linguistic assault any
longer. Can this mean that I have become too American,
that I want my nature untouched even by words?

Language

Language is the third member of the triune that gives us
a profound sense of identity and home. And the language
that does this most is one's mother tongue. My mother
tongue is Chinese. English is an acquired skill, and yet
I now speak it better. I even dream in English. The
only times I find myself using Chinese now is when I
add, subtract, or multiply. Why this should be so is a
mystery, for of course the language of arithmetic was not
something I picked up from Mother. Rather I learned it
formally at school as I had to learn English formally at
school. My attitude toward certain basic words is also
baffling. Take "mother." Not many words have deeper
emotional resonance. The Chinese word I use is *niang*.
Mother died in 1956. As the years passed, I have less and
less occasion to say the word, so when I talk about her
with my brothers I find it as natural to say "mother" as
to say *niang*. Another word I am ambivalent about is my
own name. I am so accustomed to hearing it slightly
mispronounced that when I hear it correctly pronounced,
as I did in China, I sit up wondering whether that can be
me. And then there is the standard word used to answer
the telephone: "hello" in English and *wei* in Chinese. In
China, I picked up the phone and said "wei," but I felt

self-conscious about it, as I never do when I say "hello"
or "hi" in America. But, on the other hand, for me to
say "hello" in China would be out of the question.

With practice I can regain a competence in Chinese.
I can reintegrate myself into a community—go home
again to my linguistic roots if I want to. But it will
mean accepting a small vocabulary and talking
inconsequentially of "this and that" in a narrow world.
With diligent study, I just might acquire a large
vocabulary and a rich syntax in Chinese. But then I will
have become a different person. I belatedly see what
linguists have always known: that language is a store-
house of past usages and events. When I speak English,
I am, without realizing it, an amalgam of Shakespeare,
the authors of the Book of Common Prayer, Dryden,
Osborne, and the Beatles—using their words and
phrases, which themselves reflect the times these writers
lived in. To be a truly literate and articulate Chinese, I
would have to be a different amalgam, a voice that draws
on a host of antecedent Chinese voices. It is too late for
me to do that now even if I wanted to.

So who am I? I am a citizen of the United States,
a native of China, and a human being without firm
anchorage in history, geography, and language. Like
many moderns, I feel "the unbearable lightness of being"
and have wished in weaker moments that I were more
rooted in a particular place, society, and culture. In
weaker moments, I say, for I know full well the twin
banes of rootedness, ignorance and bigotry, and for me
also this woe, that in my brief moment on earth I have
failed to use my senses and mind to the full extent that
circumstances allow.

A Pleasant Dream

What has the trip to China done for me? It has opened my eyes to the possibility of happiness. Since I have always lived the soft, pampered life of the professional class, I have known contentment and many individual satisfactions. I have even known moments of joy, seduced by the smile of a summer night, swept up by the crescendo of a Beethoven symphony. But happiness as something more intense than contentment and more lasting than joy escapes me, for it requires grounding in total acceptance, intimate ties, and a love that does not need to be repeatedly won. In our secular world, the ideal family and the ideal marriage provide these goods. I have never married. If I have the affection of friends, it is because I have worked hard to earn it, fully aware that friends will always give priority to spouse, child, and cat, and that their affection for me cools the moment I grow slack in my wooing.

I went to China expecting indifference or rejection because I abandoned it and took up citizenship elsewhere. Instead, I found a concern for my well-being that went beyond good manners, coming from people in all walks of life. That, rather than the large changes in landscape, was the shock I experienced first and foremost in China. Landscapes and cities, however beautiful or strikingly altered, will soon retreat to the back drawers of my mind. Not, however, the cab driver who, upon learning my age, told me to watch out for the curb, the tourist guide who kept telling me to drink milk and stand up straight, and the student who said I could rest my head on his shoulder if I needed to take a nap.

I was a tourist in China. I even went to the Great Wall! But I was also a native who had come home. That feeling of at homeness and of belonging is captured by the quite irrational belief that somehow I am responsible for China's past and present, that I have the native son's right to judge. In the United States, should I be ill treated, I might say to myself, "Well I am a sojourner—at best, only a naturalized citizen." Should I be ill treated in China, I can see myself becoming indignant as though my own family has turned against me. Are Americans my family, or are Chinese my family? I can't say. It is perhaps a silly question. But this thought has occurred to me. In America, I think of the students I have grown fond of as my "grandchildren." In China, the students I have grown fond of may well *be* my grandchildren.

Since my return to Madison, friends, seeing how much I have benefited from my trip, assume that I would want to go back to China, not to live there (I am too Americanized for that), but to visit. I say no, I have no wish to do so. For since my return and even when I was still in China, I have come to see that what happened to me in those two and a half weeks was unreal, the result of a unique conflation of circumstances that can never be repeated. But the trip could also seem unreal in the sense that anything too consistently pleasant could feel like a dream. I would almost rather that there were darker moments to give my experience an air of depth, a greater claim to universality. But that's not to be. Fate, up to this late point in my life, continues to smile far more often than it frowns, and never more so than on this trip to China.

I have never understood the expressions "pleasant

dreams," or "have a pleasant dream," for, in my experience, the dream, no matter how pleasant, is seldom a match for the broad-daylight realities of reading a good book, basking in the communal warmth of a coffee shop, exchanging ideas and experiences with the young. So how shall I conclude? That the China trip is just the latest and, in some ways, the rosiest installment of a long life that, in its totality, is little more than a pleasant dream?

Acknowledgments

W ithout my Wisconsin colleague A-Xing Zhu's encouragement, I would never have undertaken this trip to China, thus ending my life in the United States without taking one more look at the country of my birth. Not only A-Xing, but also his wife Ouya and their two young sons, Samuel and Alex, were also a constant source of support and delight

It is impossible for me to acknowledge all the other people who have extended to me courtesy and kindness. Many names escape me, such is the debility of old age. What follows is a bare list. I give the names of the people roughly in the order I met them: Professor Robert Sack, my colleague at Wisconsin; Professor Xing Ruan of the University of New South Wales; his student and assistant Mr. Min-chia Young; Dr. Huang Juzheng, editor of the *Architect;* Vice President Keren He of Saning Architectural Design; Dr. Xiao Ping of Liaoning University; President Shi Pei Jun, Professor Zhou Shangyi, Dean Xie Yung, and Lecturer Liu Leng Xin, all of Beijing Normal University; Deputy-director Li Xiu Bin and Dr. Zhou Chenhu of the Institute of Geographical Sciences and Natural Resources; Dr. Li Ping of the Commercial Press; Professors Yu Xixian and Zhou Yi-Xing of Peking University; Director Gao Songfan of Geographical Research in the Chinese Academy of Sciences; Director Hui Lin of the Chinese University of

Hong Kong; Dr. Tao Hiaofeng, Mr. Zhi Cheng, and Ms. Zuo Yi-Ou, currently graduate students at Beijing Normal University; Mr. Song Pu, principal of Nankai Middle School; Dean Li Ching of Chongqing University; Dean Gang Zeng of East China University; and Mr. Qiguang, currently a PhD candidate at the University of Wisconsin.

I owe a special debt to the four principal organizers and sponsors of my trip: Professors A-Xing Zhu, Xing Ruan, Zhou Shangyi, and Li Xiu Bin.

It gives me great pleasure to thank Mary Byers for improving the text and the University of Minnesota Press for publishing a work that might well be called "What I Did Last Summer"! Dare I hope that my effort will stir the creative juices of future generations of schoolchildren?

Yi-Fu Tuan is the author of *Space and Place: The Perspective of Experience, Cosmos and Hearth: A Cosmopolite's Viewpoint,* and *Dear Colleague: Common and Uncommon Observations,* all published by the University of Minnesota Press. He is professor emeritus of the University of Wisconsin–Madison, where he was J. K. Wright and Vilas Professor of Geography. He is widely considered the founder of humanistic geography.